Workbook to accompa...

Saunders

Core Concepts in Physics
CD-ROM

Workbook by

Brooke Pridmore, Ph.D.
Clayton College & State University

with problems by

Ray Serway, Ph.D.

For technical support, call 1-800-447-9457.
from 7am to 6pm CST
http://www.hbtechsupport.com

Text typeface: Berkeley Medium

Cover and text design: Jennifer Dunn

Production:
Alexandra Albin
Doris Bruey
Alexandra Buczek
Kate Davey
Paul Graham
Pat Harman
Bruce Hoffman
Samson Jarso
Sherrill Meaney
Diane Southworth
York Production Services

Printed in the United States of America
ISBN 0-03-020038-5
789 021 987654321

CONTENTS

INTRODUCTION

Saunders Core Concepts in Physics for Macintosh® and Windows™ is an interactive, three-disc CD-ROM presentation of introductory, calculus-based physics for college and university students.

FEATURES OF THE CD-ROM

The presentation:

- Uses live video, animation, interactive graphics, audio, and text to teach fundamental principles of introductory physics.
- Applies the presented concepts to real world phenomena.
- Bridges physical principles to the mathematics that describe them.
- Provides tools for learning and doing physics.

Our goal is to help you develop a deep and practical understanding of physical phenomena, to directly assist you in your study of physics. Because problem-solving is such an essential skill for success in physics, we have also included worked problems and "pop questions" within the presentation.

This three-disc set contains 14 modules (similar to chapters in a textbook) and is accompanied by this workbook, which can be used in addition to any other general physics text.

These discs also include:

- Tools such as a Unit Converter and Physical Constants table
- Notebook function
- Active indexes of contents and equations
- Worked Sample Problems and Pop Questions

**INTERACTIVE
PRESENTATION**

Information is presented as "screens" within modules. Each screen in a module introduces a key concept or a set of related concepts. Complete instructions for using this presentation are in the section of this workbook entitled *Using the CD-ROM*, and in the instructions that accompany the discs.

WORKBOOK

The workbook is organized around the main concept screens, each of which is numbered by the module and screen. The questions in the workbook can be answered after reading the text on the screen (and related screens), viewing the media, and working through associated problems.

USER'S GUIDE

To use the applications included on the CD, your system must meet the following requirements:

Macintosh installation requires:

- Macintosh computer running System 7.0.1 or greater
- 4 megabytes of free RAM minimum after system loads (5 megabytes preferred)
- 2 megabytes available hard disk space
- 256 color display at 640 × 480 pixels (minimum)
- Double-speed CD-ROM drive (minimum)

Windows installation requires:

- 486DX-33 computer running Windows 3.1 or greater
- 8 megabytes of RAM or greater
- 256 color capable VGA video board that is MPC Level II compliant, and a color VGA display (minimum)
- SoundBlaster™ compatible audio card, and speakers or a headset
- 3 megabytes available hard disk space
- Double-speed CD-ROM drive (minimum)

Installation of the *Saunders Core Concepts in Physics* presentation can be performed from any disc in the three-disc set. Only one installation is needed. Each disc contains both Macintosh and Windows versions of the presentation.

Installation Procedure for Macintosh

Note: Installation is necessary only if your system does not already have QuickTime™ version 2.0 or later. If it does, no installation is necessary. Saunders Core Concepts in Physics installs version 2.5 of QuickTime. (QuickTime is Apple Computer's programming "architecture" for motion pictures.)

Put the disc in the CD-ROM drive and double-click the *Core Concepts* disc icon on the desktop to open the CD-ROM window.

In the CD-ROM window, double-click the QuickTime Installer icon.

The installation window appears with the "Standard Installation" option pre-selected.

Note: If you are using System 7.5.0 or an earlier version, you will need Sound Manager.
To install Sound Manager, press and hold down the shift key while selecting "Sound Manager."

Click the Install button. The window indicating that the installation was successful appears. Select "Quit." (QuickTime will be placed in the Extensions folder, which is found in your System folder.)

Installation is now complete. You may need to restart your computer to run the presentation.

Note: To run the Saunders Core Concepts in Physics *presentation, one of the discs must be in the CD-ROM drive. Double-click on the Physics icon.*

Installation Procedure for Windows

Note: Installation is necessary only if your system does not already have QuickTime™ version 2.03 (or later) for Windows. If it does, no installation is necessary. Saunders Core Concepts in Physics installs version 2.1.2 of QuickTime for Windows. (QuickTime is Apple Computer's programming "architecture" for motion pictures.)

Put the disc in the CD-ROM drive.

Windows '95 and NT 4.0 Systems
Click on the "Start" button in the left corner of the Windows 95 taskbar.

Select "Run" from the pop-up menu. Type **D:\QT16.EXE** and choose "OK."

*Note: If your CD-ROM drive is not accessed through the drive letter "D,"
substitute the appropriate letter in the setup command.*

After reading the Software License Agreement, select "Agree" to continue the installation. Select "Install" from the Begin Install dialogue box.

From the Check Existing Versions dialogue box, select "Start." To complete installation, select "Install" from the Complete Install dialogue box. (QuickTime files will be placed in the System directory, which is found in your Windows directory.)

Once files have been installed successfully, select "Play Sample" for sample movie. Then select "Exit" from the File menu to exit the Movie Player.

Installation is now complete.

Note: To run the Saunders Core Concepts in Physics *presentation, one of the discs must be in the CD-ROM drive. Double-click on the Physics icon.*

Windows 3.1 or NT 3.5 Systems

From Program Manager, select "Run" from the file menu. Type **D:\QT16.EXE** and choose "OK."

*Note: If your CD-ROM drive is not accessed through the drive letter "D,"
substitute the appropriate letter in the setup command.*

After reading the Software License Agreement, select "Agree" to continue the installation. Select "Install" from the Begin Install window.

From the Check Existing Versions window, select "Start." To complete installation, select Install from the Complete Install window. (QuickTime files will be placed in the System directory, which is found in your Windows directory.)

Once files have been installed successfully, select "Play Sample" for sample movie. Then select "Exit" from the file menu to exit the Movie Player.

Installation is now complete.

Note: To run the Saunders Core Concepts in Physics *presentation, one of the discs must be in the CD-ROM drive. Double-click on the Physics icon.*

The *Saunders Core Concepts in Physics* CD-ROM is a complete multimedia presentation of college-level calculus-based introductory physics.

Core Concepts Disc 1

Starting the Presentation for Macintosh

Once the program is properly installed and the desired disc is in the CD-ROM drive, open the CD-ROM by double-clicking on its icon.

Physics

To begin the presentation, double-click on the *Physics* icon.

Starting the Presentation for Windows 95 or NT 4.0 Systems

Once the program is properly installed and the desired disc is in the CD-ROM drive, open "My Computer" and double-click on the CD-ROM icon.

Physics.exe

To begin the presentation, double-click on the *Physics.exe* icon.

Starting the Presentation for Windows 3.1 or NT 3.5 Systems

From Program Manager, select "Run" from the file menu. Type **D:\PHYSICS.EXE** and choose "OK."

> Note: If your CD-ROM drive is not accessed through the drive letter "D," substitute the appropriate letter in the setup command.

Using the Presentation

A title screen appears with a "Production Credits" bar at the screen's lower right corner. To view credits, move hand cursor and click on the bar. To view the *Contents* screen, click anywhere else on the title screen or credits screen.

The mouse is used for all navigation. Navigation within the presentation is accomplished by a single click of the mouse.

A pointing finger cursor 👆 indicates an active area.

Inactive screen areas are indicated by the arrow cursor 🡤.

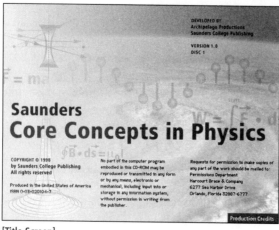

[Title Screen]

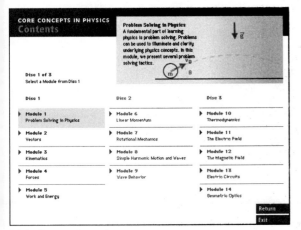

[Main Contents Screen]

The presentation is divided into modules, with Modules 1–5 on Disc 1, Modules 6–9 on Disc 2, and Modules 10–14 on Disc 3.

Click on the desired module to launch it from the *Contents* screen.

In the Contents, all active material is highlighted; all inactive material is dimmed. The material from discs other than the one in use can be accessed only by inserting the desired disc, although synopses for all modules are viewable on each disc by rolling the mouse over the module title.

The first screen of every module is a *Contents and Introduction* screen. From this screen, view the introductory "touchstone" movie or click on a topic name to open the desired Main Screen.

Each module is organized into a series of Main Screens, which address a single topic or a group of closely related topics. The *Contents and Introduction* screen provides a list of that module's Main Screens, as well as the introductory movie.

[Contents and Introduction Screen]

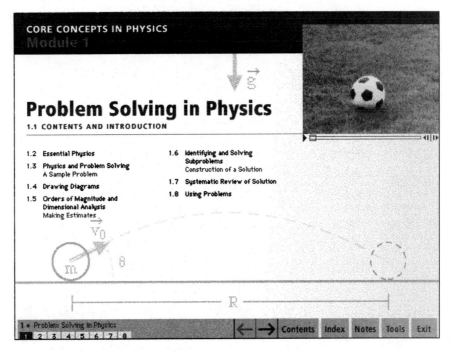

The Saunders Core Concepts in Physics CD-ROM

Navigation Bar Operation

At the bottom of the screen is the navigation bar.

The number of the screen currently in use is highlighted.

In the navigation bar, the arrow pointing left (the Back arrow) allows the user to return to the previous Main Screen. The arrow pointing right (the Forward arrow) takes the user to the next Main Screen.

The Index button accesses the index of important terms and concepts used throughout the presentation.

Selecting the Tools button displays a pop-up list of interactive tools.

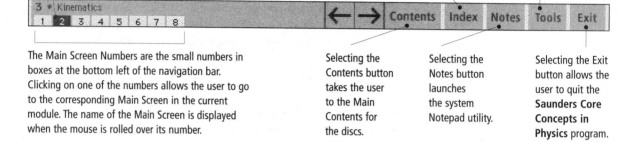

The Main Screen Numbers are the small numbers in boxes at the bottom left of the navigation bar. Clicking on one of the numbers allows the user to go to the corresponding Main Screen in the current module. The name of the Main Screen is displayed when the mouse is rolled over its number.

Selecting the Contents button takes the user to the Main Contents for the discs.

Selecting the Notes button launches the system Notepad utility.

Selecting the Exit button allows the user to quit the **Saunders Core Concepts in Physics** program.

Main Screens

Main Screens are accessed either from the Table of Contents on the Module *Contents and Introduction* screen or from the Navigation Bar at the bottom of each screen in a Module. Each Main Screen includes several features such as video or audio clips, problems, sidebars, tables, math-in-detail banners, or animated simulations that provide information about the current topic.

Many Main Screens consist of two sections. The second section can be accessed by selecting a colored bar in a corner of the screen.

Most features can be accessed and navigated by clicking on their corresponding colored arrows.

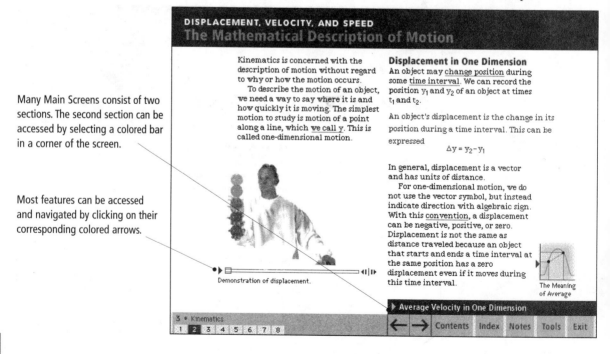

Saunders Core Concepts in Physics Workbook

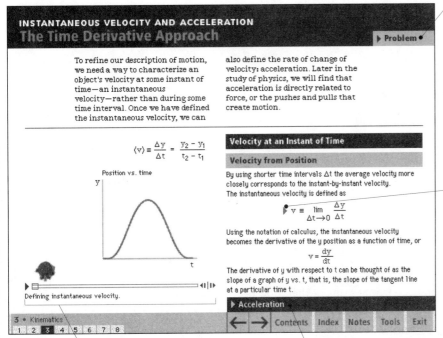

The Problem button is found in the top right corner of some screens. Select it to access a problem related to the current topic.

The arrows in the screen are used to access more information or to initiate an action such as playing an animation, or displaying mathematics-in-detail banners.

At the lower edge of most movies is a sliding control bar with play/pause button, which allows the user to see the movie with narration. The slider can be manipulated using the mouse to move forward or backward through the clip.

A Section Bar is often found at the bottom right of Main Screens; selecting it accesses the next section of the Main Screen.

Pop-up Questions are indicated by the Q icon on many screens. Select the icon for the question. To view answer choices, select the "A" in the question box. The correct answer appears after you select your answer choice.

Place the cursor over any underlined text to access a definition or explanation of that term.

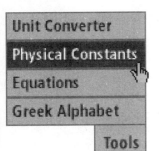

[Tools button]

Accessing the Tools Menu

Interactive tools are available from every Presentation Screen, accessed from the Tools button in the Navigation Bar. Selecting and holding the button displays a pop-up list of the tools: the Unit Converter, Equations index, Physical Constants and Physical Data tables, and more. To access a tool, roll over its name and release.

To return to the presentation screen from the tool, click on the return bar at the bottom right of the menu.

user's Guide

Reference Section and Troubleshooting Guide

This section is organized by:

- Navigation Functions
- Media Access Functions
- Utilities

Navigation Functions

[Forward/Backward Arrows]

Forward and reverse arrows allow the user to move to the next or previous Main Screen.

> *Note: The Module Opener contains no active reverse arrow, and no active forward arrow appears on the last screen of a Module.*

[Contents button]

The Contents button allows the user to return to the *Contents* screen. It also allows the user to move from one Module to another on the active disc.

At the bottom of every Module screen is a sequence of numbers corresponding to the Main Screens of that Module. The current screen is highlighted. Moving the mouse over these numbers displays the name of each Main Screen. Clicking on a number takes the user to that Main Screen.

[Main Screen Numbers]

[Exit button]

The Exit button quits the presentation.

> *Note: The* Saunders Core Concepts in Physics *program can also be terminated by choosing Quit from the file menu.*

Media Access Functions

[Colored Arrows]

Small colored arrows, found throughout the *Saunders Core Concepts in Physics* presentation, initiate some action such as accessing mathematics-in-detail banners, sidebars, or other features.

[Video Controller]

‖

[Pause]

Video buttons allow the user to play, stop, pause, and replay video or animation clips. Most clips have a sliding controller that allows users to advance or reverse the clip.

> *Note: If the video plays poorly or drops frames, or if the audio tends to "cut-out," make sure you have quit any open applications to free up additional RAM.*
> *If the problem persists, try running the application on a computer with more memory or a faster CD-ROM drive.*

Utilities
The Navigation Bar also includes the following utilities.

[Notebook access button]

The Notebook utility allows the user to enter, save, and print notes while using the *Saunders Core Concepts in Physics* program. The utility uses the standard Notepad application provided with Apple Macintosh or Microsoft Windows operating systems.

> Note: Instructions for use of the Notebook utility are available in the documentation for the Notepad function in the operating system user's guide.

If clicking on the Notes button does nothing, verify that the Notepad utility is installed on the system.

On Macintosh computers, the contents of the Notepad file are saved automatically. The file can be saved and renamed as a separate document, if desired.

In Windows, the first time you access the Notepad you will be prompted to create a file called "Physnote.txt." Choose "Yes." To keep your notes from session to session, you must save them. The file can be saved and renamed as a separate document, if desired.

[Index button]

The Index of important terms and concepts allows the user to find any important topic or term featured within the presentation. Select the colored numbers to move to a reference on the current disc.

[Index Screen]

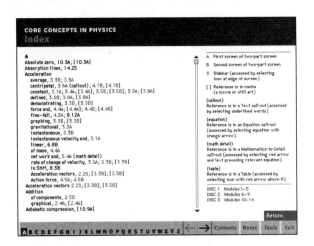

Problem Solving in Physics

It has been said that practice is the best instructor. In the study of physics, "practice" means solving problems. This workbook is designed to help you get practice.

Before tackling any physics problem, you will find it useful to equip yourself with a set of mental tools. We strongly recommend that you view the *Problem Solving in Physics* module as a first step in acquiring these tools, and to continue to refer back to it as you progress in your studies.

Here is a checklist of reminders for solving most of the problems you are likely to encounter in your physics course:

- Draw a diagram
- Estimate order of magnitude
- Perform dimensional analysis
- Identify and solve subproblems
- Construct a final answer
- Systematically review your answer
  - Check order of magnitude
  - Check units/dimensions
  - Interpret your equations for "sanity"
  - Check special cases

You will very likely wish to adapt and build on these tools as you develop your own strategies and tactics for solving problems. A key point to remember is that learning physics does not demand memorizing innumerable equations; instead, it calls for the building of bridges between various concepts and principles in physics.

Vectors

Physical phenomena can often be quantized by magnitude alone (e.g., the temperature at a given point in space). Other times, they can be specified by both a direction and a magnitude (e.g., the wind's speed and direction at the same point). Quantities that have both magnitude and direction are described by vectors. In this module, we introduce vectors and some of their applications and operations (such as addition, subtraction, and multiplication). We also discuss polar and cartesian coordinate systems, and why choosing a particular coordinate system can simplify an analysis.

DEFINITIONS

We use the three trigonometric functions:

$$\sin \theta = \frac{\text{opposite}}{\text{hypotenuse}} \qquad \cos \theta = \frac{\text{adjacent}}{\text{hypotenuse}} \qquad \tan \theta = \frac{\text{opposite}}{\text{adjacent}}$$

and the Pythagorean theorem:

$$(\text{hypotenuse})^2 = (\text{adjacent})^2 + (\text{opposite})^2$$

Vector components.

If $\vec{A} + \vec{B} = \vec{C}$, then $A_x + B_x = C_x$ and $A_y + B_y = C_y$

Scalar (dot) product.

$\vec{A} \cdot \vec{B} = |AB| \cos \theta$, where θ = angle between $\vec{A}$ and $\vec{B}$

$\vec{A} \cdot \vec{B} = A_x B_x + A_y B_y + A_z B_z$

Vector (cross) product.

$|\vec{A} \times \vec{B}| = |AB| \sin \theta$, where θ = angle between $\vec{A}$ and $\vec{B}$

Saunders Core Concepts in Physics Workbook

Right-hand rule. Orient the right hand such that the fingers curl in the direction of $\vec{A}$ toward $\vec{B}$ through the smallest angle. Lift the thumb of the right hand; it will point in the direction of $\vec{A} \times \vec{B}$.

$$\vec{A} \times \vec{B} = \begin{vmatrix} \hat{i} & \hat{j} & \hat{k} \\ A_x & A_y & A_z \\ B_x & B_y & B_z \end{vmatrix}$$

$$\vec{A} \times \vec{B} = (A_yB_z - A_zB_y)\hat{i} - (A_xB_z - A_zB_x)\hat{j} + (A_xB_y - A_yB_x)\hat{k}$$

Coordinate Systems

Problem Description

Two points in a plane have polar coordinates $(r, \theta) = (2.50$ m, $30.0°)$ and $(3.80$ m, $120.0°)$, respectively. Determine the cartesian coordinates of these points and the distance between them.

Before we begin...

1. Draw a diagram indicating the two vectors as $\vec{A}$ and $\vec{B}$.

2. How are the cartesian coordinates x and y related to the polar coordinates, r and θ?

Solving the problem

3. For each vector, $\vec{A}$ and $\vec{B}$, find the x and y components.

4. The vector separating the two points is $\vec{B} - \vec{A}$. Find the x and y coordinates of $\vec{B} - \vec{A}$.

5. Use the Pythagorean theorem to find the distance, which is the magnitude of $\vec{B} - \vec{A}$.

Vector Addition and Subtraction

Problem Description

A force $\vec{F_1}$ of magnitude 6.00 units acts on an object at the origin in a direction 30.0° above the positive x axis. A second force $\vec{F_2}$ of magnitude 5.00 units acts on the same object in the direction of the positive y axis. Use a graph to determine the magnitude and direction of the resultant force $\vec{F_1} + \vec{F_2}$.

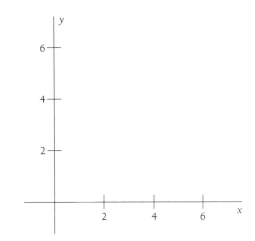

Before we begin...   1. Draw the forces from the origin in the diagram shown here.

Solving the problem   2. Given the forces as drawn, use graphical addition to find the resultant force.

Vector Components and Unit Vectors

Problem Description

A displacement vector lying in the *xy* plane has a magnitude of 50.0 m and is directed at an angle of 120.0° above the positive *x* axis. What are the rectangular components of this vector?

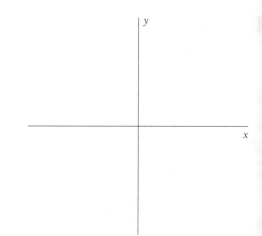

Before we begin...

1. The vector can be expressed in polar coordinate form as $\vec{A}$ = (50.0 m, 120.0°). How are the rectangular coordinates of a vector related to the polar coordinates?

2. Draw the vector on the above axis system. Drop perpendicular lines from the tip of the vector to the *x* axis and to the *y* axis. These projections give the length of A_x and A_y, respectively.

Solving the problem

3. Use the information in the above answer to find the rectangular coordinates A_x and A_y.

Vector Components and Unit Vectors

Problem Description
Instructions for finding a buried treasure
include the following: Go 75 paces at
240°, turn to 135° and walk 125 paces,
then travel 100 paces at 160°. Determine
the resultant displacement from the
starting point.

Problems

Before we begin...
1. Draw a sketch of the problem, using the graphical method of adding
vectors.

2. Express the three displacement vectors $\vec{A}$, $\vec{B}$, and $\vec{C}$ in polar coordinates.

$$\vec{A} = \qquad \vec{B} = \qquad \vec{C} =$$

Solving the problem
3. Find the x and y components of the three vectors.

(continued on next page . . .)

4. Add the *x* components together to get the *x* component of the total displacement. Do the same for the *y* components.

5. Use the Pythagorean theorem to find the magnitude of the resultant vector.

6. Use a suitable trigonometric function to determine the angle that the resultant vector makes with respect to the *x* axis.

PROBLEM 5 (SCREEN 2.6)

The Dot Product

Problem Description

Vector $\vec{A}$ extends from the origin to a point having polar coordinates (7, 70°) and vector $\vec{B}$ extends from the origin to a point having polar coordinates (4, 130°). Find $\vec{A} \cdot \vec{B}$.

Before we begin...

1. Sketch the vectors $\vec{A}$ and $\vec{B}$ on the coordinate system above. Indicate the angle that is formed between the two vectors.

2. What are the magnitudes of the two vectors?

   $|\vec{A}| =$ $|\vec{B}| =$

3. Is $\vec{A} \cdot \vec{B}$ a vector or a scalar quantity?

Solving the problem

$\vec{A} \cdot \vec{B}$ is the scalar product of the two vectors. In this problem, the vectors are expressed in polar coordinates. The scalar product is calculated according to the relation:

$$\vec{A} \cdot \vec{B} = |\vec{A}||\vec{B}| \cos \theta$$

where θ is the angle between $\vec{A}$ and $\vec{B}$.

4. Use the information gathered above to compute the value.

The Dot Product

Problem Description
Vector $\vec{A}$ is 2.0 units long and points in the positive y direction. Vector $\vec{B}$ has a negative x component 5.0 units long, a positive y component 3.0 units long, and no z component. Find $\vec{A} \bullet \vec{B}$ and the angle between the vectors.

Before we begin...

1. In this problem, the vectors are expressed in a different form than in the previous example. What form does this problem use?

2. Identify the x, y and z components of each vector:

   $A_x =$ $A_y =$ $A_z =$

   $B_x =$ $B_y =$ $B_z =$

Solving the problem

3. The scalar product in rectilinear coordinates (x, y, z) is computed according to the relationship:

   $$\vec{A} \bullet \vec{B} = A_xB_x + A_yB_y + A_zB_z$$

   Compute the scalar product.

   Notice that the scalar product is a quantity without direction. It is not a vector. If you know the scalar product, you can use the relationship:

   $$\vec{A} \bullet \vec{B} = |A||B| \cos \theta$$

   to find θ, the angle between the two. In order to find θ, what do you need to compute before substituting into the relationship?

4. Calculate the angle θ.

The Cross Product

Problems

Problem Description

Given $\vec{M} = 6\hat{i} + 2\hat{j} - \hat{k}$ and
$\vec{N} = 2\hat{i} - \hat{j} - 3\hat{k}$, calculate $\vec{M} \times \vec{N}$.

Before we begin...

1. Identify the components of the two vectors:

$M_x =$ $\qquad$ $M_y =$ $\qquad$ $M_z =$

$N_x =$ $\qquad$ $N_y =$ $\qquad$ $N_z =$

2. Is the resultant of this multiplication going to yield a vector or a scalar quantity?

Solving the problem

3. Because we have the (x, y, z) components of the two vectors, the relationship resulting from evaluating the determinant:

$$\vec{M} \times \vec{N} = \begin{vmatrix} \hat{i} & \hat{j} & \hat{k} \\ M_x & M_y & M_z \\ N_x & N_y & N_z \end{vmatrix}$$

$$\vec{M} \times \vec{N} = (M_y N_z - M_z N_y)\hat{i} - (M_x N_z - M_z N_x)\hat{j} + (M_x N_y - M_y N_x)\hat{k}$$

is used to compute $\vec{M} \times \vec{N}$.

The Cross Product

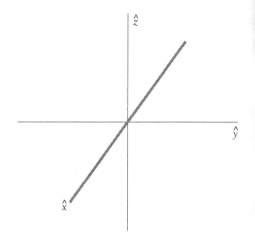

Problem Description

Vector $\vec{A}$ is in the negative y direction and vector $\vec{B}$ is in the negative x direction. What is the direction of $\vec{A} \times \vec{B}$? What is the direction of $\vec{B} \times \vec{A}$?

Before we begin...

1. Sketch the vectors on the above coordinate system. Do not worry about the lengths—only the directions are important. Why?

2. What is the relationship between the vector product of two vectors and the plane formed by the two vectors?

Solving the problem

3. Using the definition of the vector product

$$\vec{A} \times \vec{B} = |A|\,|B|\,\sin\theta$$

where θ is the angle between $\vec{A}$ and $\vec{B}$, determine the value of θ as measured counterclockwise (as per the right-hand rule).

4. If the angle is 90°, $\sin\theta$ is $+1$; if the angle is 270°, $\sin\theta$ is -1. What does this tell you about the direction of $\vec{A} \times \vec{B}$?

MODULE 3   **Kinematics**

In this module, we describe and quantify the motion of objects in one and two dimensions. This is a necessary first step before we can hope to learn about the causes of motion, which we begin to explore in Module 4, *Forces*.

The concepts used to describe motion, such as displacement, velocity, and acceleration, are vectors. For motion in one direction (i.e., back and forth along a straight-line path), a simple plus or minus sign indicates the direction of these vectors. For two- and three-dimensional motion, we use the full set of vector operations we studied in Module 2, *Vectors*.

DEFINITIONS

Displacement. The change in position of an object represented by Δx in one-dimensional motion and $\Delta \vec{r}$ in more than one dimension. It is a vector quantity that is computed by

$$\Delta \vec{r} = \vec{r}_f - \vec{r}_i$$

Velocity, $\vec{v}$. The rate of change of displacement with respect to time. The instantaneous velocity can be computed by

$$\vec{v} = \frac{d\vec{r}}{dt}$$

On a graph of displacement as a function of time, the slope of the tangent line to the displacement at the time gives the instantaneous velocity.

Acceleration, $\vec{a}$. The rate of change of velocity with respect to time.

$$\vec{a} = \frac{d\vec{v}}{dt} = \frac{d^2\vec{r}}{dt^2}$$

(*Note:* In one dimension we use x or y rather than r for the position coordinate.)

$$\vec{r} = r_x\hat{i} + r_y\hat{j}$$

For constant acceleration

$$\vec{v} = \vec{v}_0 + \vec{a}t$$

$$\vec{r} = \vec{r}_0 + \vec{v}_0t + \tfrac{1}{2}\vec{a}t^2$$

$$\vec{v}^2 = \vec{v}_0^2 + 2\vec{a}\cdot\vec{r}$$

For uniform circular motion

$$\vec{a}_c = v^2/r$$

toward the center of the circle.

For nonuniform circular motion

$$\vec{a} = \vec{a}_t + \vec{a}_r$$

where $\vec{a}_t$ is the tangential component of the acceleration.

Displacement, Velocity, and Speed

Problem Description

The velocity of a particle as a function of time is shown. At $t = 0$, the particle is at $x = 0$. Sketch the acceleration as a function of time. Determine the average acceleration of the particle from time $t = 2.0$ s to $t = 8.0$ s. Determine the instantaneous acceleration of the particle at $t = 4.0$ s.

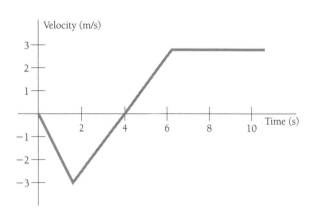

Before we begin...

The graph of velocity as a function of time given in this problem consists of three straight line segments, from (0 s, 0 m/s) to (2 s, −3 m/s) to (6 s, 3 m/s) to (8 s, 3 m/s).

1. How is the acceleration vector related to velocity vs. time?

Solving the problem

Because the average acceleration is equal to the rate of change of the velocity, we can use the relation

$$\langle \vec{a} \rangle = \Delta \vec{v} / \Delta t$$

to find the average acceleration over any time interval.

2. Does the acceleration change in the time between $t = 0$ and $t = 2$ s? What is the value of the acceleration?

3. Does the acceleration change in the time between $t = 2$ and $t = 6$ s? What is the value of the acceleration?

(continued on next page . . .)

Module 3   **Kinematics**   25

4. Does the acceleration change in the time between $t = 6$ and $t = 8$ s?

   What is the value of the acceleration?

5. You can now graph acceleration vs. time, using the answers to the above questions.

6. The average acceleration between any two time values is given by $\langle \vec{a} \rangle = \Delta \vec{v}/\Delta t$. Use the data from the original graph to answer questions relating to average acceleration.

7. Use the slope of the tangent line to the velocity vs. time graph to evaluate the acceleration at a particular point in time.

Instantaneous Velocity and Acceleration

Problem Description
The position of a softball tossed vertically upward is described by the equation $y = 7.00t - 4.90t^2$, where y is in meters and t in seconds. Find the ball's initial speed v_0 at $t = 0$. What is its velocity at $t = 1.26$ s? What is its acceleration?

Before we begin...

1. How are instantaneous velocity and acceleration related to displacement as a function of time?

Solving the problem

2. To find the initial speed of the ball at $t = 0$, we first find the velocity as a function of time. This will then be evaluated by substituting $t = 0$ into the equation. The resulting magnitude is the speed.

3. Evaluate the velocity at $t = 1.26$ s. In this portion of the problem, the sign of the answer is important. It indicates the direction of the velocity.

4. Evaluate the acceleration, again using calculus.

One-Dimensional Motion at Constant Acceleration

Problem Description

A hockey player is standing on a frozen pond when an opposing player skates by with the puck, moving with a uniform speed of 12.0 m/s. After 3.00 s, the first player makes up his mind to chase his opponent. If the first player accelerates uniformly at 4.00 m/s², how long does it take him to catch his opponent? How far has the first player traveled during this time?

Before we begin...

This problem involves describing the motion of two different objects (players). One player, called the first, undergoes an acceleration while the second player moves at constant speed.

1. When the first player catches the second, what can be said about the distance that each player has traveled? (In particular, who will have traveled farther?)

2. Identify the given information concerning the motion of each player:

First Player	Second Player
$v_{01} =$	$v_{02} =$
$a_1 =$	$a_2 =$

3. When the first player catches the second, how much time, t_2, has the second player been moving? (Assume the first player has been moving t_1 seconds.)

Solving the problem   When we examine the given information, there are two unknown quantities. Both players will have traveled the same distance, with the second one having taken 3.0 s more than the first.

4. Write an expression for the distance traveled by the first player in terms of original velocity v_{01}, acceleration a_1, and time t_1.

5. Write an expression for the distance traveled by the second player in terms of his average velocity and time t_2.

6. Since the distances traveled must be the same, the two equations can be set equal to each other.

    With the relationship between the two times established, solve for the time required for the first player to overtake the second.

7. Using either the time for the first or second player, solve for the distance that player has traveled. The two values should be the same.

Projectile Motion

Problem Description

A ball is thrown horizontally from the top of a building 35 m high. The ball strikes the ground 80 m from the base of the building. Find the time the ball is in flight. What was its initial velocity? Find the x and y components of velocity just before the ball strikes the ground.

Before we begin...

1. Draw a diagram of the situation, indicating the positive direction for both horizontal and vertical motion.

2. Using the convention that is consistent with your drawing, identify the following given information:

Horizontal	Vertical
$\langle v_x \rangle = v_{0x} =$	$v_{0y} =$
$a_x =$	$a_y =$
$\Delta x =$	$\Delta y =$

3. Since the effect of air resistance is ignored, does the horizontal velocity change?

Solving the problem

Two concepts must be applied to this problem: (1) constant acceleration in one dimension (for the vertical problem) and (2) constant velocity in one dimension (for the horizontal problem). Note that both motions occur simultaneously and therefore share the same time of flight.

4. Identify an equation of motion that relates time t, to original velocity v_{0y}, acceleration a_y, and displacement Δy.

5. Substitute the values into the equation and solve for the time of flight.

6. Now that the time of flight, the acceleration, and the initial vertical component of velocity are known, compute the final vertical component of velocity.

7. Use the horizontal displacement of the ball and the time of flight from above to calculate the horizontal component of the velocity.

Two-Dimensional Motion

Problem Description

A rifle is aimed horizontally at the center of a large target 200 m away. The initial speed of the bullet is 500 m/s. Where does the bullet strike the target? To hit the center of the target, the barrel must be at an angle θ above the line of sight. Find the angle of elevation of the barrel.

Before we begin...

1. Draw a sketch for the first part of the problem.

2. Draw a sketch for the second part of the problem.

3. Letting $+y$ be vertically up, identify the known information:

Horizontal	Vertical
$v_{0x} =$	$v_{0y} =$
$a_x =$	$a_y = -g =$
$\Delta x =$	

Solving the problem

4. Both horizontal and vertical motion take place at the same time. The time of travel will be the same for both components of the motion. Use the horizontal information to compute the time Δt required to reach the target.

5. For the vertical motion of the bullet, select one of the kinematic equations that relates the known information (including time) to the displacement. Substitute into the equation to solve for Δy.

6. In solving for the adjusted angle θ so that the bullet hits the center of the target, use the range over a level surface equation

$$R = \left(\frac{v_0^2}{g} \right) \sin (2\theta)$$

and solve for θ.

Uniform Circular Motion

Problem Description
A ball on the end of a string is whirled around in a horizontal circle of radius 0.30 m. The plane of the circle is 1.2 m above the ground. The string breaks and the ball lands 2.0 m away from the point on the ground directly beneath the ball's location when the string breaks. Find the centripetal acceleration of the ball during its circular motion.

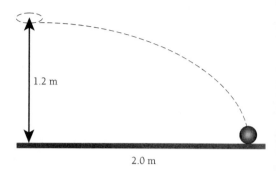

1.2 m

2.0 m

Before we begin...

1. This problem involves two different kinds of motion. What are they?

2. What two quantities must we know in order to compute centripetal acceleration?

3. What term, when computed from the projectile motion, will be used in computing the centripetal acceleration?

4. Identify the given information for projectile motion:

   $\Delta y =$ $\Delta x =$ $a_y =$ $v_{0y} =$

Solving the problem

When the string breaks, the ball moves tangentially away from the point and undergoes projectile motion. The initial velocity of the projectile motion will be the ball's velocity when the string breaks. Because the circular motion is confined to a horizontal plane, the initial velocity will be in the horizontal direction.

5. Using the projectile motion information for the problem, select an equation of motion that relates Δy, v_{0y}, a_y, and t.

6. Substitute into the equation and solve for the time required for the ball to cover the vertical displacement.

7. Using the time of flight, calculate the x component of velocity required to cover the x component of the displacement.

8. The ball's horizontal speed is the same just before and just after its release. Use this fact to calculate the centripetal acceleration of the ball.

Nonuniform Circular Motion

Problem Description

A point on a rotating turntable 20.0 cm from the center accelerates from rest to 0.700 m/s in 1.75 s. At $t = 1.25$ s, find the magnitude and direction of the centripetal acceleration and the tangential acceleration. What is the total acceleration of the point?

Before we begin...

1. What is the angle between the centripetal acceleration and the tangential acceleration at any point in time?

2. Sketch, not to scale, these two vectors and indicate their sum. Assume the angle will be reported with respect to the direction of the centripetal acceleration.

3. Identify the given information for the tangential motion of the point in the problem:

   $v_0 =$ $v_f =$ $\Delta t =$

Solving the problem

In the problem, we assume that the tangential acceleration is constant. The concepts of uniformly accelerated motion (for tangential components) and centripetal acceleration will be employed.

4. Identify the equation of motion for uniformly accelerated tangential motion that relates original velocity, final velocity, time, and acceleration.

5. Substitute into the equation and solve for the tangential acceleration, a_t.

6. Because the time in question is 1.25 seconds into the acceleration, we need to know the tangential velocity of the point at that time. Identify the equation that relates velocity, original velocity, acceleration, and time.

   Use $t = 1.25$ s, not $t = 1.75$ s. Why?

7. Substitute into the equation to evaluate the tangential velocity at that time.

8. Because the tangential velocity and the radius are known, evaluate the centripetal acceleration.

9. Because a_r and a_t are at right angles to one another, calculate their vector sum using the Pythagorean theorem.

Relative Motion

Problem Description

A boat crosses a river of width $w = 160$ m in which the current has a uniform speed of 1.50 m/s. The pilot maintains a bearing (i.e., the direction in which the boat points) perpendicular to the river and a throttle setting to give a constant speed of 2.00 m/s relative to the water. What is the speed of the boat relative to a stationary shore observer? How far downstream from the initial position is the boat when it reaches the opposite shore?

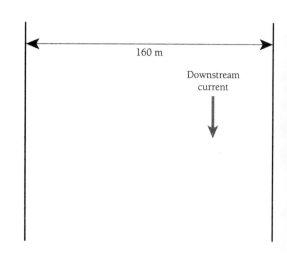

160 m

Downstream
current

Before we begin...

1. In a diagram, show the velocity of the boat relative to the river, $\vec{v}_{br}$, and the velocity of the river relative to the shore, $\vec{v}_{rs}$. Indicate the vector sum $\vec{v}_{bs}$ of the two velocities. (Ignore the scale of the width of the river.)

2. Why is the vector sum important to this problem?

Solving the problem

Not only is the boat moving across the river due to the boat's motor, but it is moving downstream due to the river's current. A stationary observer on the shore will see the relative velocity due to both of these influences. In solving this problem, the concepts of vector addition and uniform motion will be employed.

3. Use the Pythagorean theorem to calculate the magnitude of the velocity of the boat relative to the shore $\vec{v}_{bs}$. This is the speed of the boat relative to the shore observer.

4. Since the heading of the boat is directly perpendicular to the current, the boat speed relative to the water will determine the time required for the crossing. Assuming that the river is perpendicular to the x axis, the boat will move across the river with $\vec{v}_{br} = 2.00$ m/s. Calculate the time required to cross the river.

5. The entire time the boat is moving across the river, it is also moving downstream at the rate of 1.50 m/s. Use the appropriate equation of motion to calculate the distance downstream that the boat will travel.

MODULE 4   **Forces**

In the previous module, we learned to describe motion elegantly and concisely. In this module, we begin to explore the forces that cause this motion. Newton's three laws of motion relate the net force acting on an object to its resulting motion. No force is necessary to keep a moving object in motion, but it does take a net force to make any object accelerate.

Because forces have both magnitude and direction, they are vector quantities.

DEFINITIONS

Newton's first law. An object at rest remains at rest and an object in motion continues in motion, at constant velocity, unless acted upon by a net external force.

Newton's second law. The net force acting on an object at any instant is equal to the object's mass multiplied by its instantaneous acceleration.

Newton's third law. When two bodies interact, the force exerted by body 1 on body 2 is equal in magnitude and opposite in direction to the force exerted by body 2 on body 1.

Weight. The force exerted on an object by the Earth's gravitational attraction.

Normal force. The force that occurs between objects in direct contact, which keeps them from falling through each other. The normal force is always directed perpendicular (normal) to the surface of contact.

Centripetal force. The force that accelerates a mass into circular motion.

Newton's second law expressed in equation form is

$$\vec{F}_{net} = M\vec{a}$$

Weight is computed as mass times acceleration due to gravity:

$$\vec{w} = M\vec{g}$$

The equation of motion resulting from Newton's second law is

$$\sum \vec{F} = m\frac{d^2\vec{x}}{dt^2}$$

Newton's third law can be expressed in equation form by

$$\vec{F}_{12} = -\vec{F}_{21}$$

Centripetal force is calculated by

$$F_r = ma_r = m\frac{v^2}{r}$$

Problems

Motion, Newton's First Law, and Force

Problem Description

Two people are pulling a boat through the water as shown in the figure. Each person exerts a force of 600 N directed at a 30.0° angle relative to the forward motion of the boat. If the boat moves with constant velocity, find the resistive force, $\vec{F}$, exerted on the boat by the water.

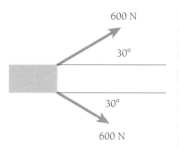

600 N

30°

30°

600 N

Before we begin...

1. If the boat moves at a constant velocity, what is the net force that is being exerted on the boat?

2. Label the forces that the people are exerting as $\vec{T}_1$ for the uppermost force and $\vec{T}_2$ for the lower oriented force. Draw the resistive force onto the diagram shown here.

Solving the problem

3. Because the boat is not accelerating, Newton's first law tells us that the sum of the forces acting upon the boat must equal zero. Selecting the boat to be a point mass at the origin of the cartesian coordinate system, resolve the forces into their x and y components. The x components must sum to zero, as must the y components.

4. Solve for the x and y components of the resistive force $\vec{F}$.

Inertia, Mass, and Weight

Problem Description

If the Earth's gravitational force causes a falling 60-kg student to accelerate downward at 9.80 m/s², determine the upward acceleration of the Earth during the student's fall. Take the mass of the Earth to be 5.98×10^{24} kg.

Before we begin...

1. What is the relationship between net force, mass, and acceleration?

Solving the problem

2. According to Newton's third law, the student exerts a force of exactly the same magnitude, but opposite in direction, on the Earth. Compute the magnitude of the force exerted by the Earth on the student (the student's weight).

3. Set this force equal to the Earth's mass multiplied by the acceleration and solve for the value of the acceleration.

Newton's Second Law

Problem Description

A constant force changes the speed of an 85-kg sprinter from 3.0 m/s to 4.0 m/s in 0.50 s. Calculate the magnitude of the acceleration of the sprinter, the magnitude of the force, and the magnitude of the acceleration of a 58-kg sprinter experiencing the same force. (Assume linear motion.)

Before we begin...

1. Identify the given information for this problem:

$$\vec{v}_i = \qquad \vec{v}_f = \qquad \Delta t = \qquad m_1 = \qquad m_2 =$$

2. What quantities are to be computed?

Solving the problem

3. Using the given information from above and the equations for uniform acceleration from Module 3, compute the acceleration that the sprinter must experience.

4. If the acceleration of a given mass is known, which one of Newton's laws of motion allows you to calculate the net force acting upon the mass?

5. Use this information to find the force on the sprinter.

6. Finally, use this known force to calculate the unknown acceleration of the 58-kg sprinter.

Newton's Third Law

Problem Description

Three blocks are in contact with each other on a frictionless, horizontal surface. A horizontal force $\vec{F}$ is applied to M_1. If $M_1 = 2.00$ kg, $M_2 = 3.00$ kg, $M_3 = 4.00$ kg, and $F = 18.0$ N, draw free-body diagrams of each block and find the acceleration of the blocks. What is the *resultant* force on each block? What are the magnitudes of the contact forces between the blocks?

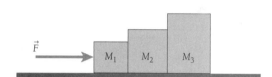

Before we begin...

1. Draw the free-body diagrams for the three blocks.

2. What is the relationship of the acceleration of each block to the accelerations of the other blocks?

Solving the problem

3. Using the free-body diagrams, apply Newton's second law to relate the net force on each block to the block's mass and acceleration. Take care to keep the appropriate signs to the forces.

4. Solve the three simultaneous equations for the acceleration.

5. Substitute the value of the acceleration into each of the second law equations to evaluate the contact force acting upon the object. (*Hint:* Begin with either Block 1 or Block 3, not Block 2.)

Free-Body Diagrams

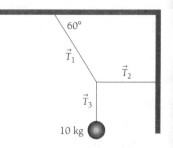

Problem Description

Find the tension in each cord of the system illustrated here.

Before we begin...

1. Draw two free-body diagrams for the problem. Include all forces acting upon the junction of the string in the first diagram and all forces acting upon the ball in the second.

2. Is the system in equilibrium, or does a net force act upon the object?

3. What is the weight of the ball?

Solving the problem

4. Because the system is in equilibrium, the net force acting upon the ball must be zero. From the second free-body diagram, compute the magnitude of the tension, $\vec{T}_3$.

5. The first free-body diagram shows that the tension $\vec{T}_3$ is pulling vertically downward ($-y$ direction). What are the directions of $\vec{T}_1$ and $\vec{T}_2$?

   Since the knot is in equilibrium, $\vec{T}_1 + \vec{T}_2 + \vec{T}_3 = 0$. This requires that simultaneously:

   $$T_{1x} + T_{2x} + T_{3x} = 0 \quad \text{and} \quad T_{1y} + T_{2y} + T_{3y} = 0$$

6. Resolve the tensions into their x and y components and solve the two equations simultaneously.

Free-Body Diagrams

Problem Description

Three objects are connected by strings as shown in the figure. The masses are 4.0 kg, 1.0 kg, and 2.0 kg, respectively, and both the table and the pulleys are friction-less. Determine the acceleration of each object and their directions. Determine the tensions in the two cords.

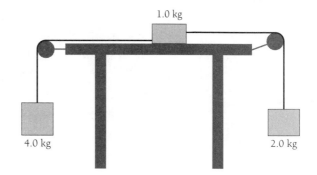

1.0 kg

4.0 kg   2.0 kg

Before we begin...

1. Which object is most likely to accelerate in the vertically downward direction, the 4.0-kg object or the 2.0-kg object?

2. What is the relationship between the accelerations of the objects?

3. Draw free-body diagrams for each of the objects. Consider the positive direction for each object to be the direction you have predicted it will move.

Solving the problem

4. From the information above and the free-body diagrams, write Newton's second law as it applies to each of the three objects.

5. Solve the equations simultaneously to find the acceleration of the system. This must be the same value for each of the objects.

6. Substitute back into the several equations to evaluate the tensions in the two cords.

Centripetal Force

Problem Description
An automobile moves at a constant speed over the crest of a hill. The driver moves in a vertical circle of radius 18.0 m. At the top of the hill, she notices that she barely remains in contact with the seat. Find the speed of the vehicle.

Before we begin...	1. Draw a free-body diagram for the driver as she crests the hill. Does a normal force exist between the driver and the seat?
	2. What force is providing the centripetal acceleration that she is experiencing?
Solving the problem	3. Since the driver is moving at a constant speed in a vertical circle, she must have a centripetal force acting toward the center of the circle. Write the expression for the net force providing the centripetal force and set it equal to centripetal force.
	4. Solve this for the speed of the car.
	5. Why does the mass of the driver not matter in this problem?

Fictitious Forces: Motion in Accelerated Reference Frames

Problem Description

A 5.00-kg mass attached to a spring scale rests on a frictionless, horizontal surface as in the figure. The spring scale, attached to the front end of a boxcar, reads 18.0 N when the car is in motion. If the scale reads zero when the car is at rest, determine the acceleration of the car while it is in motion. What will the spring scale read if the car moves with constant velocity? Describe the forces on the mass as observed by someone in the car and by someone at rest outside the car.

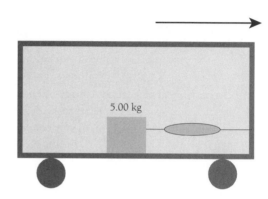

5.00 kg

Before we begin...

1. Draw a free-body diagram for the mass as observed from the ground. Assume that the spring scale exerts a tension $\vec{T}$ directed in the $+x$ direction upon the mass; this tension is measured by the scale.

Solving the problem

2. The boxcar is accelerating in the $+x$ direction. Since there is no friction, the net force exerted on the mass m is the tension $\vec{T}$. Calculate the acceleration of the mass with respect to the inertial frame of reference (a stationary point on the ground.)

3. If the car moves with constant velocity, what is its acceleration? Use this information to calculate the tension under this condition.

4. In the noninertial frame of reference (within the boxcar), the mass appears to be in equilibrium. What is the net force acting upon an object in equilibrium? Use this information to compute the fictitious force in the accelerated frame of reference.

Problems

MODULE 5   **Work and Energy**

INTRODUCTION

The concepts of work and energy and the law of conservation of energy are important in examining mechanical systems. These concepts build upon those of Module 4, *Forces* (and Newton's laws of motion). Using the work-energy relationships often simplifies the analysis of a situation, in contrast to a direct application of Newton's second law. Work, energy, and energy conservation are important concepts in all areas of physics.

In this module, we use the scalar product of two vectors to derive a scalar quantity.

DEFINITIONS

Work. The force applied to an object through a distance.

Kinetic energy. The energy that an object has by virtue of its motion.

Conservative force. The work done by a force that is independent of the path taken from the start to the finish position.

Potential energy. The energy that an object has by virtue of its state, shape, or position.

Power. A measurement of the rate at which work is done or energy is used.

Conservation of energy. A law stating that energy cannot be created or destroyed. Energy can only be converted among the three forms of energy: kinetic, potential, and thermal energy.

The work done by a constant force is computed by the equation

$$W = \vec{F} \cdot \vec{s}$$

where $\vec{F}$ is the applied force and $\vec{s}$ is the displacement.

If the force varies over the displacement, use the equation

$$W = \int_{start}^{finish} \vec{F} \cdot d\vec{s}$$

The work-energy theorem can be expressed by the equation

$$W_{net} = \Delta K$$

where K is the kinetic energy ($^1/_2 mv^2$).

Potential energy U is defined by the equation

$$\Delta U = -W_C = -\int_{start}^{finish} \vec{F}_C \cdot d\vec{s}$$

Power is computed by the equation

$$P = \frac{dW}{dt} = \vec{F} \cdot \vec{v}$$

Work

Problem Description

A force $\vec{F} = (4.0x\hat{i} + 3.0y\hat{j})$ N acts on an object that moves in the x direction from the origin to $x = 5.0$ m. Find the work done on the object by the force.

Before we begin...

The force has units of newtons N, and the displacement has units of meters m. The numeric constants 4.0 and 3.0 respectively in the force equation must have units of N/m.

1. How is the work done by a variable force calculated?

2. What is the direction of the displacement $\vec{ds}$ in this problem?

Solving the problem

3. The force is applied over a displacement from $x = 0$ to $x = 5.0$ m. These become the limits of integration for evaluating the work. Recall the definition of the scalar (dot) product from Module 2. Using the answers to questions 1 and 2 above, write and evaluate the integral expression for the work done.

Important Examples of Work: Gravity and Springs

Problem Description
A cheerleader lifts his 50.0-kg partner straight up off the ground a distance of 0.60 m before releasing her. If he does this 20 times, how much work has he done?

Before we begin...

1. How much force does the cheerleader have to exert to lift his partner each time?

2. Through what displacement is this average force exerted each time?

Solving the problem

3. Employ the equation for work done by a constant force to calculate the work done for each lift.

4. Because the process was done 20 times, multiply the work done for a single lift by 20.

Important Examples of Work

Problem Description

When an ideal spring is stretched, the spring force satisfies the equation $\vec{F} = -kx$. If the spring constant k is 10 N/m, calculate the work done by this force when the spring is stretched 0.10 m.

Before we begin...

Since the force is not constant, the work will have to be computed using the relation

$$\int \vec{F} \cdot d\vec{x}$$

1. What will be the limits of the integral?

Solving the problem

2. Evaluate the integral between the limits; substitute the value of k into the expression.

Work Done to Accelerate a Mass

Problem Description

A 100-g bullet is fired from a rifle having a barrel 0.6 m long. Assuming the origin is placed where the bullet begins to move, the force (in newtons) exerted on the bullet by the expanding gas is in the positive-x direction, with magnitude $15{,}000 + 10{,}000x - 25{,}000x^2$, where x is in meters. Determine the work done by the gas on the bullet as the bullet travels the length of the barrel. If the barrel is 1.00 m long, how much work is done? How does this value compare with the work calculated in the first case?

Before we begin...

1. There are three terms in the force equation. Each term must have units of newtons. Evaluate the units of the coefficients of each term.

2. What is the expression for calculating the work done by a variable force?

Solving the problem

3. Substitute the given force into the equation for determining work done by a variable force. Notice that the force is in the x direction. This simplifies $d\vec{s}$ to $d\vec{x}$; θ is $0°$.

4. Evaluate the integral between the limits of $x = 0$ and $x = 0.6$ m.

5. Repeat between the limits of $x = 0$ and $x = 1.00$ m. Compare the results.

Conservative Forces

Problem Description

A single conservative force acting on a particle varies as $\vec{F} = (-Ax + Bx^2)\hat{i}$ N, where A and B are constants and x is in meters. Calculate the potential energy associated with this force, taking $U = 0$ at $x = 0$. Find the change in potential energy and change in kinetic energy as the particle moves from $x = 2.0$ m to $x = 3.0$ m.

Before we begin...   1. What are the concepts that will apply to this problem?

Solving the problem   2. Because the work done by a conservative force equals the negative of the change in potential energy, you can evaluate the potential energy function by integrating the force over the interval of the displacement.

3. The conservation of energy principle tells us that $\Delta K + \Delta U = 0$. Evaluate the work done in going from $x = 2.0$ m to $x = 3.0$ m. Use this information to answer the second question in the problem.

Work-Energy Theorem

Problem Description

A 15.0-g bullet is accelerated in a rifle barrel 72.0 cm long to a speed of 780 m/s. Use the work-energy theorem to find the average force exerted on the bullet while it is being accelerated.

Before we begin...

1. Identify the given information:

   $m =$ $\vec{v}_0 =$ $\vec{v}_f =$

2. State the work-energy theorem.

3. Over what distance does the force accelerate the bullet?

Solving the problem

4. The work-energy theorem allows us to equate the work done by the net force to the change in kinetic energy. Compute the change in kinetic energy of the bullet. (Be careful with the units.)

5. Since the average force acts over the entire distance of the barrel, use the definition of work to calculate the average force.

Problems

Power

Problem Description
A 650-kg elevator starts from rest.
It moves upward for 3.00 s with constant
acceleration until it reaches its cruising
speed, 1.75 m/s. What is the average
power of the elevator motor during this
period? How does this power compare
with its power while it moves at its
cruising speed?

Before we begin...

1. Draw a free-body diagram of the forces acting upon the elevator.

2. Identify the given information:

$$m = \qquad v_0 = \qquad v_f = \qquad \Delta t =$$

3. During the 3.00 seconds, the work done by the elevator motor is going to change what two types of energy for the elevator?

Solving the problem

4. Since both kinetic energy and potential energy are changed during the acceleration interval, the work done by the motor during this time equals the sum of these changes. We know the speeds, so the change in kinetic energy can be computed.

5. To compute the change in potential energy, we need to know the height Δy to which the elevator rises in 3.00 s. Use the kinematic equations to calculate this height. After finding Δy, compute the change in potential energy.

6. Calculate the average power as the work done by the motor divided by the time required.

    For objects moving at constant speed, power can be calculated by the relation

    $$\langle P \rangle = \vec{F} \bullet \langle \vec{v} \rangle$$

7. What force is required to keep the elevator moving at a constant speed?

    Now, evaluate the cruising speed power requirement.

Conservation of Energy

Problem Description
At time t_i, the kinetic energy of a particle is 30 J and its potential energy is 10 J. At some later time t_f, its kinetic energy is 18 J. If only conservative forces act on the particle, what are its potential energy and its total energy at time t_f? If the potential energy at time t_f is 5 J, are there any nonconservative forces acting on the particle? Explain.

Before we begin...

1. State the law of conservation of energy.

2. Identify the given information for the first question:

$K_i =$ $\Sigma U_i =$ $K_f =$

3. For the second question, what is ΣU_f?

Solving the problem

4. Applying the law of conservation of energy for a conservative system, compute the potential energy at t_f.

5. Evaluate the total energy E at time t_i. This is the total energy at t_f as well.

6. The condition under which a nonconservative force acts upon the system is when there is a change in the total kinetic energy plus potential energy of the system. Has this occurred in the second question in this problem?

MODULE 6 | Linear Momentum

In this module, we introduce the law of conservation of linear momentum. We study the general form of Newton's second law of motion as well as collisions between objects. Finally, we learn to express the motion of a system of objects in terms of the center of mass of the system.

Linear momentum, $\vec{p}$. The mass of an object multiplied by its velocity. Linear momentum is a vector quantity.

General form of Newton's second law. The net force acting upon a system is equal to the rate at which the linear momentum of the system changes with time.

Impulse, I. The change in momentum of the system. The product of the force and the time for which it acts.

Elastic collision. A process that conserves linear momentum and mechanical energy.

Inelastic collision. A process that conserves linear momentum but does not conserve mechanical energy.

Center of mass. A point that, in terms of mechanical behavior, would move as the system moves, if the point were to contain the entire mass of the system. The linear momentum of the center of mass is the sum of the linear momenta of all objects in the system.

The equation for computing the linear momentum of a particle is

$$\vec{p} = m\vec{v}$$

If the net force applied to a system is zero, then linear momentum is conserved and constant.

$$\Delta\vec{p}_T = 0$$

The general form of Newton's second law of motion is

$$\vec{F}_{net} = \frac{d\vec{p}}{dt} = \sum_i \left(m_i\frac{d\vec{v}_i}{dt} + \vec{v}_i\frac{dm_i}{dt} \right)$$

The impulse of a system is computed by the equation

$$I = \int_{t_1}^{t_2} \vec{F}(t)dt = \Delta\vec{p} = \langle\vec{F}\rangle_T \Delta t$$

For a system of particles, the center of mass r_{CM} is calculated by the equation

$$\vec{r}_{CM} = \frac{\sum m_i\vec{r}_i}{M}$$

For a continuous distribution of particles, such as an extended object, the center of mass is found from the evaluation of the equation

$$\vec{r}_{CM} = \frac{1}{M}\int\vec{r}\,dm$$

The momentum of the system can be expressed in terms of the center of mass by the relationship

$$\vec{p}_{total} = M\vec{v}_{CM} = M\frac{d\vec{r}_{CM}}{dt}$$

The acceleration of the center of mass is given by Newton's second law:

$$\vec{a}_{CM} = \frac{\sum\vec{F}_{net}}{M}$$

The General Form of Newton's Second Law

Problem Description
A rocket engine consumes 80 kg of fuel per second. If the exhaust speed is 2.5×10^3 m/s, calculate the thrust on the rocket.

Before we begin...

1. State the general form of Newton's second law.

2. How is thrust computed?

3. Identify the given information:

$$\frac{dM_g}{dt} = \qquad\qquad\qquad v_g =$$

Solving the problem

Newton's second law tells us that the total force exerted upon a system is equal to the external forces acting upon it and the thrust due to changing mass with respect to time. In this problem, the external force acting upon the rocket is not required because we are only interested in the thrust.

4. Write the expression for thrust and substitute the given information to evaluate the thrust.

Impulse

Problem Description

The force F_x in Newtons acting on a 2.0-kg particle varies in time as shown. Find the impulse of the force, the final velocity of the particle if it is initially at rest, and its final velocity if it is initially moving along the x axis with a velocity of −2.0 m/s. What is the average force exerted on the particle for the time interval $t_i = 0$ to $t_f = 5.0$ s?

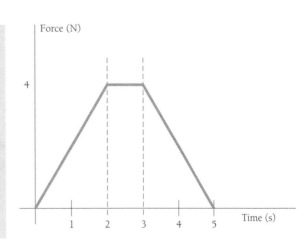

Before we begin...

1. Identify the given information:

   $m =$ $v_{i(\text{particle initially at rest})} =$ $v_{i(\text{particle initially moving})} =$

2. How is impulse related to force and the time interval over which the force is applied?

3. Recall that impulse equals the change in momentum. How is impulse related to mass and velocity?

Solving the problem

4. Since the force is not constant over the entire interval, you will need to employ the relation

   $$\Delta \vec{p} = \sum (F_i \Delta t_i)$$

   for each of the three regions (two triangles and a rectangle) indicated in the graph.

(continued on next page . . .)

Module 6   **Linear Momentum**   65

5. Write the equation for final momentum in terms of the initial momentum and the impulse.

6. Evaluate the equation for the given information about mass and initial velocity.

7. Show how large a constant force instead of a variable force would be required to act over the same 5 s time period in order to give the same value of impulse according to the relation

$$\Delta \vec{p} = \langle \vec{F} \rangle \Delta t$$

Perfectly Inelastic Collisions

Problem Description

A 90-kg halfback running north with a speed of 10 m/s is tackled by a 120-kg opponent running south with a speed of 4.0 m/s. If the collision is perfectly inelastic and head-on, calculate the speed and direction of the players just after the tackle and determine the energy lost as a result of the collision. Account for the missing energy.

Before we begin...

1. Draw a sketch, using blocks to represent the two players. Label the masses and the velocities before the collision. Assign the positive direction.

2. Identify the given information:

   $m_1 =$ $m_2 =$ $\vec{v}_1 =$ $\vec{v}_2 =$

3. What is meant by a perfectly inelastic collision?

4. Is momentum conserved in this collision?

Solving the problem

5. To solve the problem in which momentum is conserved but mechanical energy is not, evaluate the total momentum of the system before the collision.

(continued on next page . . .)

6. Because the momentum after the collision is the same as before the collision, the velocity of the two players (who stick together and move as one mass) can now be evaluated.

7. Evaluate the total kinetic energy before the collision and compare it with the total afterward. Are they the same? Why or why not?

Perfectly Inelastic Collisions

Problem Description
During the battle of Gettysburg, the gun-fire was so intense that several bullets collided in midair and fused together. Assume a 5.0 g Union musket ball moving to the right at 250 m/s, 20° above the horizontal, and a 3.0 g Confederate ball moving to the left at 280 m/s, 15° above the horizontal. Immediately after they fuse together, what is their velocity?

Before we begin...

1. Draw a sketch of the system just before the collision. Label the objects and assign the positive and negative directions.

2. Identify the given information:

$$m_1 = \qquad m_2 = \qquad \vec{v}_1 = \qquad \vec{v}_2 =$$

3. What kind of collision do the two bullets undergo?

4. Is this problem solved in one or in two dimensions?

Solving the problem

5. The perfectly inelastic collision takes place in two dimensions. This requires that the momentum in the x direction be conserved, as is the momentum in the y direction. Find the x and y components of momentum for each bullet before the collision.

(continued on next page . . .)

Problems

6. Set the total x momentum before the collision equal to the combined mass multiplied by the x component of velocity after the collision. Solve for $\vec{v}_x$.

7. Repeat the above procedure for the y component.

8. Knowing the x and y components of velocity, find its magnitude and direction.

Perfectly Elastic Collisions

Problem Description

Two blocks of mass m_1 = 2.00 kg and m_2 = 4.00 kg are released from a height of 5.00 m on a frictionless track as shown. The blocks undergo an elastic head-on collision. Determine the two velocities just before the collision. Determine the two velocities immediately after the collision. Determine the maximum height to which each block rises after the collision.

2 kg   4 kg

5 m   5 m

Before we begin...

1. Identify the given information:

   $m_1 =$ $m_2 =$ $\Delta h =$

   $v_1 =$ $v_2 =$

2. Considering that the track is frictionless, what happens to the gravitational potential energy of the blocks as they slide to the bottom of the incline?

3. In an elastic collision, what two quantities are conserved?

Solving the problem

4. Use the law of conservation of energy to solve for the speeds of the two blocks immediately before the collision. Assign the appropriate sign to each of these to convert them to velocities. (Let movement toward the right be +.)

(continued on next page . . .)

Problems

5. Evaluate the total linear momentum and kinetic energy before the collision. Equate these values to the expressions for total linear momentum and kinetic energy after the collision, and solve for the velocities after the collision.

6. You can now employ the law of conservation of energy for each block after the collision to compute the height to which each will rise.

Center of Mass

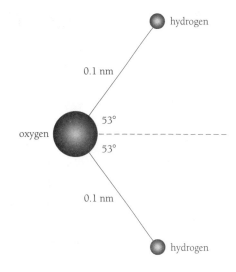

hydrogen

0.1 nm

53°

oxygen

53°

0.1 nm

hydrogen

Problem Description

A water molecule consists of an oxygen atom with two hydrogen atoms bound to it. The angle between the two bonds is 106°. If each bond is 0.100 nm long, where is the center of mass of the molecule?

Problems

Before we begin...

The mass, in grams, of a single atom is given by its mass in AMU's divided by Avogadro's number. The mass of a single atom of H and O are 1.67×10^{-24} g and 2.66×10^{-23} g, respectively.

1. Assuming that the oxygen atom in the above drawing is located at the origin and the dashed line is along the x axis, is there any symmetry that will simplify the problem?

2. What is the y component of the center of mass (y_{CM})?

Solving the problem

3. Assuming that, because of symmetry, $y_{CM} = 0$, find the x component of each hydrogen atom's position.

4. Write the equation for the location of the x_{CM} for a set of discrete particles, and substitute the values of mass and position for this problem.

Motion of a System of Particles

Problem Description

A 2.0-kg particle has a velocity of $\vec{v}_1 =$ $(2.0\hat{i} - 10t\hat{j})$ m/s, where t is in seconds. A 3.0-kg particle moves with a constant velocity of $\vec{v}_2 = 4.0\hat{i}$ m/s. At $t = 0.50$ s, find the velocity of the center of mass. What is the acceleration of the center of mass? What is the total momentum of the system?

Before we begin...

1. Identify the given information:

   $m_1 =$ $m_2 =$ $\vec{v}_1 =$ $\vec{v}_2 =$

2. How is the linear momentum of the center of mass of a system related to the total linear momentum of the system?

3. How is acceleration related to velocity as a function of time?

Solving the problem

4. Compute the total linear momentum. This will be the momentum of the center of mass of the system as well.

5. Using the formula for $\vec{v}_{CM}$, solve for the velocity of the center of mass.

6. Evaluate this function at the indicated time.

7. Use the definition of acceleration as related to velocity as a function of time to compute the acceleration of the center of mass.

8. Find the total momentum of the system, either by calculating and adding up the momentum of each particle or by the formula

$$\vec{p}_{tot} = m_{tot}\vec{v}_{CM}$$

Motion of a System of Particles

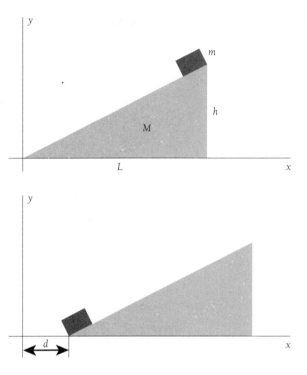

Problem Description

A block of mass $m = 2.0$ kg is placed on the top of a movable incline of mass $M = 8.0$ kg, height $h = 2.0$ m, and the base length $L = 6.0$ m. If the block is released from rest, how far will the incline have moved when the block reaches the bottom? Assume all surfaces are frictionless. (*Hint*: The x coordinate for the center of mass of the block-incline system is fixed. Why?)

Before we begin...

1. Identify the given information:

   $M =$ $m =$

   $L =$ $h =$

2. Is there an external force having an x component that acts upon the system? What does this say about $\Delta \vec{p}_{CM(x)}$ for the center of mass of the system?

Solving the problem

The system consists of two objects, each having a center of mass. The center of mass of the system is computed by taking the two centers for the objects and treating these as point masses.

3. The first step is to calculate the center of mass of the right triangle that has a length L and a height h. Consult the drawing shown at the top of the next page. Write an expression for dm in terms of the surface mass density, M/A, and the area of the strip ydx.

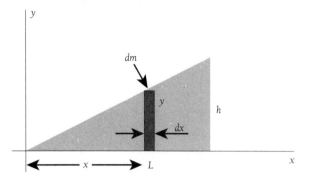

4. Evaluate the x_{CM} by the relation

$$x_{CM} = \frac{1}{M} \int x\, dm$$

5. The location of the center of mass of the total system can be calculated by placing M at the x_{CM} of the triangle and m at the position $x = L$.

6. Because the x_{CM} of the system does not change, set the value calculated equal to having mass m located at d and M located at $(x_{CM \text{ triangle}} + d)$.

MODULE 7

Rotational Mechanics

To describe the motion of a system, we must go beyond translational motion and also consider rotational motion. This module shows the analogies between translational and rotational concepts of position, velocity, and acceleration. In addition, rotational counterparts for mass and force are introduced.

Rotational work, energy, and power are discussed, as are angular momentum and its conservation.

This module requires the cross product of multiplying two vectors. Review Module 2 if you are not familiar with the vector cross product.

DEFINITIONS

Angular velocity, ω. The rate of change of angular position θ with respect to time. The direction of the angular velocity is determined by using the right-hand rule.

Angular acceleration, α. The rate of change of angular velocity with respect to time.

Moment of inertia. The rotational counterpart of mass, represented by the mass multiplied by the square of the distance.

Torque, τ. The rotational counterpart of force. Torque depends upon the applied force and the position of the applied force with respect to the point of rotation.

Angular momentum. The rotational counterpart to linear momentum. Just as linear momentum is equal to the product of mass and velocity, angular momentum L is computed by the product of moment of inertia and the angular velocity.

Law of conservation of angular momentum. In the absence of a net external torque, the angular momentum of a system remains constant.

In solving rotational kinematic problems, the equations of motion mirror the equations from translational kinematics for constant acceleration.

$$\theta = \theta_0 + \omega_0 t + \frac{1}{2}\alpha t^2$$

$$\omega = \omega_0 + \alpha t$$

$$\theta = \theta_0 + \frac{1}{2}(\omega_0 + \omega)t$$

$$\omega^2 = \omega_0^2 + 2\alpha(\theta - \theta_0)$$

The moment of inertia for a system of discrete particles is

$$I = \sum m_i r_i^2$$

where r is the perpendicular distance of the particle of mass m with respect to the axis of rotation.

The moment of inertia is calculated by the equation

$$I = \int_{body} \rho r_\perp^2 dV$$

where ρ is the mass density of the object.

The rotational kinetic energy K_R is found from the equation

$$K_R = \frac{1}{2}I\omega^2$$

The parallel axis theorem relates the moment of inertia about an axis to the moment of inertia about the center of mass.

$$I = I_{CM} + Md^2$$

where d is the distance from the center of mass to the axis of rotation.

Torque is evaluated by the relation

$$\vec{\tau} = \vec{r} \times \vec{F}$$

Work and power are computed using the equations

$$dW = \tau d\theta \quad \text{and} \quad P = \vec{\tau} \cdot \vec{\omega}$$

Angular momentum is found using the equation

$$\vec{L} = \vec{r} \times \vec{p} = I\vec{\omega}$$

Newton's second law for rotational systems is

$$\vec{\tau}_{net} = \frac{d\vec{L}}{dt}$$

Basic Concepts in Rotational Kinematics

Problem Description

A discus thrower accelerates a discus from rest to a speed of 25.0 m/s by whirling it through 1.25 rev. Assume the discus moves on the arc of a circle 1.00 m in radius. Calculate the final angular speed of the discus. Determine the magnitude of the angular acceleration of the discus, assuming it to be constant. Calculate the acceleration time.

Before we begin...

1. Identify the given information:

   $$v_i = \qquad v_f = \qquad \theta = \qquad R =$$

2. How is angular speed related to tangential or linear speed?

Solving the problem

3. Convert the initial and final speeds to angular speeds using the linear transformation equations.

4. Identify an equation from rotational kinematics that involves initial and final angular speeds, the angle through which the acceleration occurs, and the angular acceleration.

5. Before solving the problem, convert the angle from revolutions to radians.

6. Substitute into the identified equation and solve for angular acceleration.

7. To find the required time for the acceleration, select any of the rotational kinematic equations involving time and the given information, including the now calculated acceleration. Substitute and solve for the time.

Rotational Energy

Problem Description
A solid cylinder of unknown composition has a mass 215 g, length 10.8 cm, and diameter 6.38 cm. It is placed at rest on the top of an incline that is 3.00 m long and at 25° to the horizontal. Using energy methods, calculate the moment of inertia of the cylinder if it takes 1.50 s to reach the bottom of the incline. You cannot assume that the cylinder's density is uniform.

Before we begin...

1. Identify the given information:

   $m =$ $r =$ $\Delta x =$ $\theta =$

   $\Delta t =$ $v_i =$

2. Does the length of the cylinder really matter as long as it is rolling while laying on its side?

3. State the law of conservation of energy, ignoring nonconservative work done on the contents of the cylinder. (*Hint*: There are two types of kinetic energy in the problem.)

Solving the problem

4. It will be necessary to find the final translational and angular speeds of the cylinder as it reaches the bottom of the incline. Select the appropriate translational kinematics equations to find v_f and then use this value to compute ω_f. (You can assume that the linear acceleration is constant.)

5. Calculate the height of the incline in order to determine the cylinder's initial gravitational potential energy.

6. Assuming that the cylinder rolls without slipping, no work is done by friction between the cylinder and the incline. Set the total energy at the top of the incline equal to the total energy at the bottom. Since all quantities are known, except for the moment of inertia I, evaluate and solve for I.

Moment of Inertia of Rigid Bodies

Problem Description

Two masses M and m are connected by a rigid rod of length L and negligible mass, as shown. For an axis perpendicular to the rod, show that the system has the minimum moment of inertia when the axis passes through the center of mass. Show that this moment of inertia is $I = \mu L^2$ where $\mu = mM/(m + M)$.

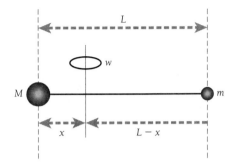

Before we begin...

1. Assuming that the rod is massless, the total moment of inertia is a result of two point masses rotating about the axis. What is the moment of inertia of a point mass rotating about a fixed axis?

Solving the problem

2. Write the equation for the total moment of inertia of the two-object system.

3. Since the moment of inertia is a function of x, the first and second derivative tests for relative maximum and minimum values can be employed. Taking the first derivative of the function and setting it equal to zero, find the values of x that satisfy the equation.

4. Evaluate whether this is a maximum or minimum value, substituting the value from the above solution into the second derivative with respect to x. If the value of the second derivative is positive, the value of x is a minimum.

5. Now evaluate the moment of inertia of the system about an axis passing through the value of x calculated in question 4. Use algebra to obtain the desired result.

Torque

Problem Description

A particle is located at the vector position $\vec{r} = (\hat{i} + 3\hat{j})$ m, and the force acting on it equals $(3\hat{i} + 2\hat{j})$ N. What is the torque about the origin and about the point having coordinates (0, 6) m?

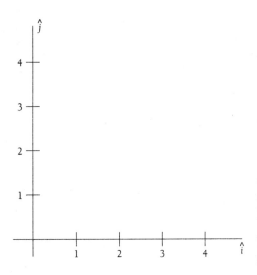

Before we begin...

1. Identify the given information in unit vector notation:

$$\vec{r} = \qquad\qquad \vec{F} =$$

2. Sketch the position vector and draw the force vector from the end of the position vector. Locate the position of the second point of rotation and draw a vector from that point to the position of the first vector.

3. How is torque related to position and applied force?

Solving the problem

This problem requires the concept of cross product (vector) multiplication from Module 2.

4. Using the determinant form, write the expression for $\tau = \vec{r} \times \vec{F}$. Solve for the torque.

5. Evaluate the position vector with respect to the new point of rotation. Write the position vector in terms of the new point of rotation.

6. Again using the determinate form, write the expression for $\tau = \vec{r} \times \vec{F}$. Use the particle location with respect to the new point of reference. Solve for the torque.

7. What do you notice about the direction of the torque? What is its relationship to the plane that contains the position and force vectors?

Work and Energy in Rotational Motion

Problem Description

A potter's wheel—a thick stone disk of radius 0.50 m and mass 100 kg—is freely rotating at 50 rev/min. The potter can stop the wheel in 6.0 s by pressing a wet rag against the rim. Find the total change in rotational kinetic energy and the net torque to bring the wheel to rest.

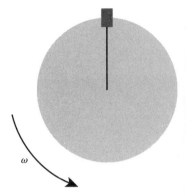

ω

Before we begin...

1. Identify the given information.

$$M = \qquad R = \qquad F_n = \qquad \Delta t = \qquad \omega_i =$$

2. The friction from the potter's rag will create a torque about the axis of rotation of the wheel. Will this torque cause the angular speed to increase or decrease? How do you know?

3. What will be the final rotational kinetic energy of the wheel, after the torque has brought the wheel to rest?

Solving the problem

The concepts to be applied to this problem are (1) the work-energy theorem; (2) rotational kinematics; and (3) the definition of torque in terms of applied force and moment arm.

4. To apply the work-energy theorem, we need to recall that the rotational kinetic energy is given by $K = \frac{1}{2}I\omega^2$. Using the equation $I = \frac{1}{2}MR^2$ for a solid disk, calculate the moment of inertia of the wheel.

5. Next, convert the angular speed, given in rev/min, to radians/second. (Why is this important?)

6. Compute the initial rotational kinetic energy and the change in kinetic energy during the problem.

7. Considering that the work done by the net torque is equal to the change in kinetic energy, write the formula for work done by a constant net torque and set it equal to the change in kinetic energy.

8. Find $\Delta\theta$. Because the angle, $\Delta\theta$, through which the torque was applied was not given, you need to apply the rotational kinematic equations with the given information.

9. Now you can evaluate and solve for the torque.

Rolling Motion

Problem Description
A uniform solid disk and a uniform hoop are placed side by side at the top of an incline of height h. If they are released from rest and roll without slipping, determine their speeds when they reach the bottom. Which object reaches the bottom first?

Before we begin...

1. Draw a sketch illustrating the problem.

2. What are the moments of inertia for the objects?

   $I_{disk} =$ $I_{hoop} =$

3. What kind of energy does each object have at the top of the incline?

Solving the problem

4. The law of conservation of energy tells us that the gravitational potential energy possessed by the object as it is released to roll down the incline will be converted to kinetic energy as it rolls. Since the object does not slip, no work is done against friction. All of the potential energy lost will result in gained kinetic energy. A rolling object has both rotational kinetic energy and translational kinetic energy. Write the expression for each type of kinetic energy.

5. Write the law of conservation of energy as it applies to this problem. Substitute the known quantities into the equation. Do this separately for the disk and the hoop.

6. Solve the resulting equations for the speed at the bottom by recalling the relationship between angular speed and translational speed for a rotating object. Which has the greater speed at the bottom?

7. The object with the greater speed at the bottom will have reached the bottom first. Why?

Angular Momentum

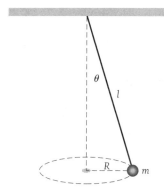

Problem Description

A conical pendulum consists of a bob of mass m moving in a circular path in a horizontal plane as shown. During the motion, the supporting wire of length l maintains a constant angle θ with the vertical. Show that the magnitude of the angular momentum of the bob about the point at the center of the circle is

$$L = \sqrt{\frac{m^2 g l^3 \sin^4 \theta}{\cos \theta}}$$

Before we begin...

1. How is the angular momentum of a particle rotating in a constant circular path computed?

2. Draw a free-body diagram for the problem.

3. If a particle is moving in a constant circular path, what must be the net force acting upon it?

Solving the problem

4. Using the information provided in the free-body diagram, apply Newton's second law for both x and y components. Equate the centripetal force ($ma_r = mv^2/R$) to the net x force.

5. Solve the two equations simultaneously to find v.

6. Substitute the expressions for R and v into the equation for angular momentum.

Conservation of Angular Momentum

Problem Description

A merry-go-round of radius $R = 2.0$ m has a moment of inertia $I = 250$ kg·m² and is rotating at 10 rev/min. A 25-kg child, who is initially at rest, steps onto the edge of the merry-go-round. What is the new angular speed of the merry-go-round?

Before we begin...

1. State the law of conservation of angular momentum.

2. How is the angular momentum related to the moment of inertia of a system?

3. Identify the given information:

   $I_{\text{m-g-r}} =$ $\omega_{\text{m-g-r}} =$

   $R =$ $m_c =$

4. What is the equation for computing the moment of inertia of a revolving point mass particle? (The child will be considered a point mass in this problem.)

Solving the problem

Because no net external torque is applied to the system (child plus merry-go-round), the angular momentum of the system will remain unchanged. Further, the total moment of inertia of the system about the point of rotation equals the sum of the individual moments of inertia of the objects in the system.

5. Evaluate the angular momentum of the merry-go-round.

6. After the child jumps onto the merry-go-round, what is the new moment of inertia of the system?

7. Use the information to solve for the new rotational speed after the child comes aboard.

MODULE 8

Simple Harmonic Motion and Waves

INTRODUCTION

Simple harmonic motion (SHM) is the periodic oscillation about a stable equilibrium point. A wave, on the other hand, can be thought of as a series of simple harmonic oscillators incrementally "out of phase" with each other.

In this module, we explore both these forms of periodic motion. We learn to describe them mathematically and to develop an understanding of the linear restoring forces that drive them.

DEFINITIONS

Amplitude, A. The maximum displacement from the equilibrium position.

Period of motion, T. The time required for one complete oscillation or cycle.

Angular frequency, ω. A constant related to the period of motion by $\omega = 2\pi/T$.

Phase constant, ϕ. An angular offset that describes where a wave or oscillation is in its cycle at time $t = 0$.

Wavelength, λ. The distance between one peak of a wave and the adjacent peak (or one trough and the adjacent trough).

Wave number, k. A number equal to 2π divided by the wavelength.

USEFUL EQUATIONS

When analyzing the motion as a result of applied forces, use Newton's second law. In SHM, we use Hooke's law:

$$F_{net} = ma = -kx$$

The general equation for solving SHM becomes

$$\frac{d^2x}{dt^2} = -\omega^2 x$$

Substituting the expression for Hooke's law into this equation gives us

$$\frac{d^2x}{dt^2} = -\left(\frac{k}{m}\right)x$$

Position as function of time for SHM is given by

$$x = A \cos(\omega t + \phi)$$

Upon solving the general equation, the angular frequency for the system is given by

$$\omega = \sqrt{\frac{k}{m}} \qquad \text{for the spring-mass system}$$

$$\omega = \sqrt{\frac{g}{l}} \qquad \text{for the simple pendulum}$$

The period of motion T and the frequency f are related to the angular frequency by

$$T = \frac{2\pi}{\omega} \qquad \text{and} \qquad f = \frac{\omega}{2\pi}$$

The general wave function for a sinusoidal wave moving to the right is of the form

$$y = A \cos(kx - \omega t + \phi)$$

where the wavelength λ is related to the wave number k by

$$\lambda = \frac{2\pi}{k}$$

The speed of a wave v is related to the wavelength and frequency by

$$v = f\lambda$$

Simple Harmonic Motion

Problem Description

A 20-g particle oscillates with a frequency of 3.0 Hz and an amplitude of 5.0 cm. Through what total distance does the particle move during one cycle of its motion? What is its maximum speed and where does it occur? Find the maximum acceleration of the particle. Where in the motion does the maximum acceleration occur?

Before we begin...

1. Does the mass of the particle matter?

2. Identify the given information:

   $f =$ $A =$

3. Draw a sketch of the position as a function of time for the particle.

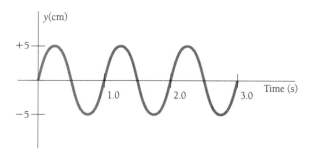

Solving the problem

4. To evaluate the total distance traveled, examine the graph you just drew. What is the distance between the equilibrium point and the maximum displacement? This distance is traveled four times. Verify this on your sketch.

5. If the equation for the displacement as a function of time is $y(t) = A \sin (2\pi ft + \phi)$, use the fact that

$$v = \frac{dy}{dt}$$

to write an expression for $v(t)$. What is the maximum value that this expression can have?

6. Use the above information and

$$a = \frac{dv}{dt} = \frac{d^2x}{dt^2}$$

to write an expression for $a(t)$. What is the maximum value that this expression can have?

Physical Nature of Waves

Problem Description

A transverse wave on a string is described by $y = (0.12 \text{ m}) \sin \pi(x/8 + 4t)$. Determine the transverse speed and acceleration of the string at $t = 0.20$ s for the point on the string located at $x = 1.6$ m. What are the wavelength, period, and speed of propagation of this wave?

Before we begin...

1. Write the general expression for a traveling wave on a string.

2. How are the transverse speed and acceleration related to the position as a function of time of a traveling wave?

Solving the problem

3. Using the relation between position and velocity, evaluate the derivative of displacement with respect to time at the given position and time.

4. Repeat the above procedure using the relation between velocity and acceleration.

5. Compare the general expression for a traveling wave to the equation given in this problem. Identify the requested values by inspection.

$$\lambda = \qquad\qquad T = \qquad\qquad v =$$

Mathematical Nature of Waves

Problem Description

A sinusoidal wave traveling in the −x direction (to the left) has an amplitude of 20.0 cm, a wavelength of 35.0 cm, and a frequency of 12.0 Hz. The displacement of the wave at $t = 0$, $x = 0$ is $y = -3.00$ cm, and the wave has a positive velocity here. Sketch the wave at $t = 0$. Find the angular wave number, period, angular frequency, and phase velocity of the wave. Write an expression for the wave function $y(x, t)$.

Before we begin...

1. Identify the given information:

$$A = \qquad \lambda = \qquad f = \qquad y(0, 0) =$$

2. Sketch the wave at $t = 0$. In making this sketch, you are taking a snapshot of the wave at $t = 0$. You are actually plotting y as a function of x. What do you have to know in order to make this plot?

3. Write the general expression for the wave traveling to the left as a function of position and time.

4. How will knowing the value of y at $t = 0$ and $x = 0$ help you find the phase constant?

Solving the problem

Since the wave has a negative displacement but a positive velocity, the graph of y vs. t must begin at -0.03 m and have a positive slope at that point. It should repeat itself every 0.35 m, having a maximum displacement of ±0.2 m.

5. Use the values of wavelength and period to identify the wave number and angular frequency.

6. Evaluate the phase constant by substituting $x = 0$ and $y = 0$ into the traveling wave equation and solving for ϕ.

7. Write the equation for this particular traveling wave using the general form and the computed values.

Mathematical Nature of Waves

Problem Description

A transverse sinusoidal wave on a string has a period $T = 25.0$ ms and travels in the negative x direction with a speed of 30.0 m/s. At $t = 0$, a particle on the string at $x = 0$ has a displacement of 2.00 cm and travels to the left with a speed of 2.00 m/s. What is the amplitude of the wave? What is the initial phase angle? What is the maximum transverse speed of the string? Write the wave function for the wave.

Before we begin...

1. What is the general wave function for a wave traveling to the left (negative x direction) as a function of position and time?

2. Identify the given information:

   $T =$ $v_x =$ $y(0, 0) =$ $v_y(0, 0) =$

3. How is the period of a wave T related to the angular frequency ω?

4. What is meant by *initial phase angle*?

Solving the problem

5. At $(0, 0)$, the wave function is reduced to $y = A \cos \phi$, and the wave speed is $dy/dt = -A\omega \sin \phi$. Because we know how to evaluate ω from the period, we can use the trig identity

$$(A \sin \theta)^2 + (A \cos \theta)^2 = A^2$$

to solve for A.

6. Substitute the calculated value of A back in the expressions for $y(0, 0)$ to yield the values of ϕ.

7. Recalling that $v_{y \, max} = A\omega$, evaluate for the maximum transverse wave speed.

8. By finding the wave number k, we can write the wave function for the wave. Recall the relationship between wave speed (not transverse speed), period, and wavelength. Use this and the relationship between wave number and wavelength to evaluate k.

9. Substitute the known quantities into the general wave function to give the specific wave function.

Mathematical Nature of Waves

Problem Description

A sinusoidal wave train is described by
$y = (0.25 \text{ m}) \sin (0.30x - 40t)$, where x and
y are in meters and t is in seconds.
Determine for this wave the amplitude,
angular frequency, wave number, wavelength, wave speed, and the direction of
the propagation.

Before we begin...

1. Notice that, in this problem there does not appear to be a phase angle. The equation is expressed in terms of a sine function rather than a cosine function. This will not affect our results. Write the general expression for the wave function of a traveling wave.

2. According to the general expression, is this wave traveling in the positive or in the negative x direction?

Solving the problem

3. Do a term-by-term comparison of the specific equation and the general equation to identify the following:

$$A = \qquad\qquad k = \qquad\qquad \omega =$$

4. Using the general wave equation, $v = f\lambda$, and the relationships between k and λ and between ω and v, find the wavelength and wave speed.

5. If the sign before the t term in the equation is positive, the motion is to the left. If the sign is negative, the motion is to the right ($+x$ direction).

Hooke's Law and the Equation of Motion

Problem Description

A 1.0-kg mass attached to a spring of force constant 25 N/m oscillates on a horizontal, frictionless track. At $t = 0$, the mass is released from rest at $x = -3.0$ cm. (That is, the spring is compressed by 3.0 cm.) Find the period of its motion, the maximum values of its speed and acceleration, and the displacement, velocity, and acceleration as functions of time.

Before we begin...

1. Identify the given variables and terms:

$$m = \qquad\qquad k = \qquad\qquad A =$$

2. What are the three subproblems to this problem?

Solving the problem

3. What is the relation between the given information and the period of a spring-mass system? Use this to evaluate T.

(continued on next page . . .)

Module 8   **Simple Harmonic Motion and Waves**   107

4. To compute the maximum speed, do the following. In finding the maximum speed of the mass, employ the law of conservation of energy, and then calculate the work done in compressing the spring from equilibrium. Since the total energy of the spring-mass system before the work is done was zero, the total energy after compressing the spring will be equal to the work done upon the string by the external agent.

The maximum speed will occur when all of the energy of the system is kinetic (potential energy will be zero).

$$K = 2mv^2 \quad \text{and} \quad W = 2kA^2$$

5. Finding the maximum acceleration requires recalling that the force exerted by the spring on the mass obeys Hooke's law. Accordingly, if Newton's second law of motion is employed, the acceleration will be a maximum when the displacement is a maximum. The direction of the acceleration will not be considered. We are asked to find its maximum value only.

$$\vec{F}_{net} = m\vec{a} \quad \text{and} \quad \vec{F} = -k\vec{x}$$

To identify the forms expressing displacement, velocity, and acceleration as functions of time, recall that the general position as a function of time for a system in SHM is

$$x = A \cos (\omega t + \phi)$$

and that

$$v(t) = \frac{dx}{dt} \quad a(t) = \frac{dv}{dt}$$

6. Substituting the given value of A and the computed value of ω into the equation for position as a function of time, evaluate the initial conditions to find ϕ. Evaluate the derivative of x with respect to t to find v. Repeat to solve for the acceleration.

SHM and Waves in the Real World

Problem Description
A simple pendulum has a length of 3.00 m. Determine the change in its period if it is taken from a point where $g = 9.80$ m/s^2 to an elevation where the free-fall acceleration decreases to 9.79 m/s^2.

Before we begin...

1. For small amplitudes, what is the relation between the period of motion of a simple pendulum, the pendulum's length, and the acceleration due to gravity?

2. How do you calculate the change in a quantity as measured under two different conditions?

Solving the problem

3. Use the equation relating the period of motion to the length of a simple pendulum and the acceleration due to gravity to calculate the periods for each of the given acceleration values.

4. Subtract the value computed for $g = 9.80$ m/s^2 from the value computed when $g = 9.79$ m/s^2.

MODULE 9   **Wave Behavior**

In Module 8, *Simple Harmonic Motion and Waves*, we introduced wave motion for traveling waves and defined the concepts of wavelength, frequency, and wave speed. In this module, we investigate how boundary conditions and other waves affect the overall propogation of a wave through a medium, as well as how the energy and power of that wave is affected by such conditions.

DEFINITIONS

Boundary. The interface between any two mediums. Waves at a boundary can be reflected and/or transmitted.

Superposition principle. When two waves combine, they pass straight through without interruption or distortion. At a position y, the total disturbance is the sum of the individual disturbances:

$$y = y_1 + y_2$$

Interference. The addition or subtraction of the amplitudes of two waves located at the same position and time yields an amplitude that is constructive if the two are in phase and destructive if the two are 180° out of phase.

Standing wave. A wave in a confined region that "oscillates in place" rather than moving across space. Mathematically, these wave patterns always result from interference between two or more moving waves.

Harmonics. Stable modes of vibration corresponding to particular frequencies. The lowest allowable frequency is called the first harmonic.

Node. A point of zero amplitude in a standing wave, created by ongoing destructive interference.

Antinode. A position of relative maximum amplitude in a standing wave created by ongoing constructive interference.

Resonance. When energy is added to a system at the system's natural frequency, the amplitude of oscillations is maximized.

The speed of a one-dimensional mechanical wave such as a pulse in a stretched string is computed by

$$v = \sqrt{\frac{F_t}{\mu}}$$

where F_t is the tension in the string and μ is the linear mass density of the string.

The energy E and the power P associated with a wave are calculated by the equations

$$E = \frac{1}{2}\mu\omega^2 A^2 L \qquad \text{and} \qquad P = \frac{1}{2}\mu\omega^2 A^2 v$$

where ω is the angular frequency of the wave, A is the wave amplitude, L is the length of the string, and v is the speed of the wave.

For standing waves, the relationships between the allowable frequencies of vibration and the length of the string are given by

$$f_n = nv/2L \qquad n = 1, 2, 3... \qquad \text{(wave fixed at both ends or free at both ends)}$$

$$f_n = nv/4L \qquad n = 1, 3, 5... \qquad \text{(wave fixed at one end and free at the other end)}$$

Speed of a Wave in a Medium

Problem Description

A light string of mass per unit length 8.00 g/m has its ends tied to two walls separated by a distance equal to 3/4 the length L of the string. A mass m is suspended from the center of the string, putting a tension in the string. Find an expression for the transverse wave speed in the string as a function of the hanging mass. How much mass should be suspended from the string to have a wave speed of 60.0 m/s?

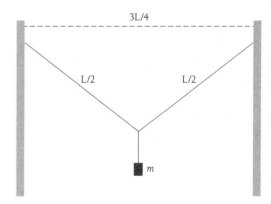

Before we begin...

1. Identify the given information:

   $\mu =$ $v =$

2. Draw the free-body diagram for the system.

3. The tension in each section of the string is the same. This ensures that the x components of the forces sum to zero. What do the y components of the tensions have to equal to ensure equilibrium?

4. How can the angle θ that the string makes with respect to the horizontal be computed?

5. The string and the mass are in equilibrium, requiring that the vector sum of the forces be zero. By symmetry, the x components of the tension will be equal in magnitude to each other. Apply the condition of equilibrium to the y components and solve for the tension T.

6. The relation between the transverse wave speed and the string through which it is traveling is

$$v = \sqrt{\frac{F_t}{\mu}}$$

where F_t is the tension and μ is the linear mass density.

The tension depends upon the weight that is supported. Solve the above given equation for tension and then substitute into the relation between tension and weight.

Energy and Power in Waves

Problem Description

Transverse waves are being generated on a rope under constant tension. By what factor is the required generating power increased or decreased if the length of the rope is doubled and the angular frequency remains constant? The power is changed by what factor when the amplitude is doubled and the angular frequency is halved?

When both the wavelength and amplitude are doubled, what happens to the power? What happens to the power when both the length of the rope and the wavelength are halved?

Before we begin...   1. Write the expression for the power delivered by a transverse wave.

2. Does the length of the rope affect the power?

Solving the problem   3. If the length changes, it does not affect the power. Isolate the dependency for the various combinations of variables described in the four questions asked in the description. Answer the questions.

Superposition and Interference

Problem Description
Two waves are traveling in the same direction along a stretched string. Each has an
amplitude of 4.0 cm, and they are 90° out
of phase. Find the amplitude of the resultant wave.

Before we begin...

1. Write the wave function of each of the two waves.

2. State the superposition principle.

Solving the problem

3. Recall the trigonometric identity

$$\sin(a) + \sin(b) = 2 \sin((a + b)/2) \cos(a - b)/2)$$

Let $a = (kx - \omega t)$ and $b = (kx - \omega t - \phi)$. In this problem, $\phi = 90°$.
Substitute and simplify the equation.

Standing Waves

Problem Description

Two waves given by $y_1(x, t) = A \sin (kx - \omega t)$ and $y_2(x, t) = A \sin (2kx + \omega t)$ interfere. Determine all x values where there are stationary nodes. Determine all x values where there are nodes that depend on time t.

Before we begin...

1. What are node positions?

2. According to the superposition principle, what must be true for a point to be a node?

Solving the problem

3. Use the superposition principle to find the wave function of the combined waves.

4. Stationary nodes occur as a result of a term that is independent of time t equaling zero. Identify the term in the combined equation, set it equal to zero, and solve for the values of x that satisfy the equation.

5. Time dependent nodes arise from a term that has both position x and time t dependency. Identify such a term in the equation and solve it by setting it equal to zero.

Standing Waves—Wave Fixed at Both Ends

Problem Description

A 2.0-m long wire having a mass of 0.10 kg is fixed at both ends. The tension in the wire is maintained at 20 N. What are the frequencies of the first three allowed modes of vibration?

Before we begin...

1. What is the relationship for the allowable wavelengths for a wave fixed at both ends?

2. How can the speed of the wave be calculated for a wire under tension?

Solving the problem

3. Compute the mass density μ for the wire and use this result to find the speed of the wave.

4. Calculate the frequencies by equating the frequency f_n from the speed and the allowable wavelengths λ_n.

Standing Waves—Wave with One Fixed End and One Free End

Problem Description

A student uses an audio oscillator of adjustable frequency to measure the depth of a water well. Two successive resonant frequencies are heard at 52.0 Hz and 60.0 Hz. What is the depth of the well?

Before we begin...

1. The water level in the well is treated as a node, and the top of the well is an antinode. In this problem, one end is fixed and one end is free. What is the relationship between the length of the system and allowable frequencies for this type of system?

2. Identify the given information:

$$f_a = \qquad\qquad f_b =$$

Solving the problem

3. Write the expression for the allowable frequency with $a = n$ and for the allowable frequency with $b =$ next allowable frequency.

4. Solve the equations simultaneously by setting the difference in frequencies $f_b - f_a$ equal to the difference in their computational formulas. This will eliminate n from the equation. We now have change in frequency as a function of wavespeed and depth of the well.

5. Use the speed $v = 344$ m/s (speed of sound at standard temperature and pressure). Solve for the depth of the well.

Resonance

Problem Description

A weight of 40.0 N is suspended from a spring that has a force constant of 200 N/n. The system is undamped and is subjected to a harmonic force of frequency 10.0 Hz, resulting in a forced-motion amplitude of 2.00 cm. Determine the maximum value of the force.

Before we begin...   1. Identify the given information:

$$m = w/g = \qquad k = \qquad A = \qquad \omega =$$

Let the harmonic force be represented by

$$F = F_0 \sin \omega t = ma_0 \sin \omega t$$

The equation of motion for the system subjected to this driving force is found by applying Newton's second law:

$$m\frac{d^2x}{dt^2} = -kx + ma_0 \sin \omega t$$

Dividing all terms by the mass m gives

$$\frac{d^2x}{dt^2} = -\frac{k}{m}x + a_0 \sin \omega t$$

Upon substituting $\omega_0^2 = k/m$, the equation of motion can be written as

$$\frac{d^2x}{dt^2} = -\omega_0^2 x + a_0 \sin \omega t$$

Solving the problem

Use the trial solution

$$x = C \sin \omega_0 t + A \sin \omega t$$

2. Evaluate the second derivative of x with respect to time t.

3. Substitute the second derivative and the trial solution value of x into the equation of motion.

4. Simplify the equation and solve for A that satisfies it.

5. The maximum value of the force occurs when all of the amplitude of the motion is in A; thus $C = 0$. The amplitude of the oscillation is equal to the absolute value of A.

   Use the given information to evaluate the acceleration a_0 resulting from the driving force. This information is then used to find the force F_0.

MODULE 10   **Thermodynamics**

The laws of thermodynamics allow us to express the relationship between thermal energy transfer, work, and internal energy. These laws also provide limits on the efficiency of thermal processes.

     The classical view of thermodynamics allows us to analyze the effects of thermal energy on the gross properties of matter. Our study centers around what happens to the system. We use the statistical mechanical approach to give insight into the thermal processes at the molecular level and then relate the properties to measurable quantities such as pressure and temperature.

Internal energy. The collective term for all forms of energy internal to a substance (not influenced by the overall translation or rotation of the body as a whole). Chemical energy, nuclear energy, and thermal energy are internal energy types.

Thermal energy. The total internal mechanical energy of the molecules of a substance.

Thermal equilibrium. The process that occurs when two bodies are placed in thermal contact, and no net heat flows between them. (Heat is another name for thermal energy transfer.)

Temperature. If two bodies placed in thermal contact remain in thermal equilibrium, then they are said to have the same temperature. If they do not, then heat will flow from the body with higher temperature to the one with lower temperature. For most types of systems, temperature is proportional to the average kinetic energy of the molecules.

Heat capacity. A measure of the amount of thermal energy that is required to raise an object's temperature by a specific amount.

Absolute zero. The lowest possible temperature. At this temperature, molecules of a substance have essentially zero thermal energy. Absolute zero corresponds to −273.15 °C.

Zeroth law of thermodynamics. Two objects that are independently in thermal equilibrium with a third object are in thermal equilibrium with each other.

First law of thermodynamics. The first law of thermodynamics is a generalization of the law of conservation of energy to include thermal energy. Any thermal energy absorbed by a system increases the system's internal energy or goes into work done by the system, or both.

Adiabatic process. A process that occurs when there is no net thermal energy transfer.

Isothermal process. A process that occurs when there is no change in temperature.

Second Law of Thermodynamics. In a closed system, the total entropy either increases or stays the same.

Entropy. A measure of the disorder of a system.

USEFUL EQUATIONS

The heat capacity C of a substance is calculated by the equation

$$C = \frac{Q}{\Delta T}$$

where Q is the thermal energy transferred and ΔT is the change in temperature.

The ideal gas law relating pressure, volume, and temperature is described by the equation

$$PV = Nk_BT$$

where N is the number of gas molecules in the sample and k_B is a constant known as Boltzmann's constant.

The first law of thermodynamics can be expressed in differential form as

$$dQ = dW + dU$$

with W representing the work done by the system and U being the internal energy.

The efficiency of a heat engine is computed by

$$e = \frac{\text{work}}{\text{energy absorbed}} = \frac{Q_h - Q_c}{Q_h}$$

where Q_h is the heat absorbed from the hot reservoir and Q_c is the heat deposited to the cold reservoir.

For a Carnot engine, the efficiency is

$$e = \frac{T_h - T_c}{T_h}$$

with the temperatures being measured on the absolute temperature scale.

The coefficient of performance (COP) of a Carnot heat pump is

$$\text{COP} = \frac{Q_h}{W} = \frac{T_h}{T_h - T_c}$$

The entropy of a system is computed by using the equation

$$dS = \frac{dQ_r}{T}$$

where dQ_r is the change in thermal energy along a reversible path.

Basic Concepts of Thermodynamics

Problem Description
A perfectly insulated calorimeter contains 500 ml of water at 30°C and 25 g of ice at 0°C. Determine the final temperature of the system.

Problems

Before we begin...

1. The system is insulated and isolated from its surroundings. It will come to thermal equilibrium. What is meant by thermal equilibrium?

The concept of heat capacity C is discussed on the CD-ROM. Associated with the concept is another concept called specific heat c. Specific heat is the amount of heat per unit mass required to change the temperature of a substance by a given amount.

$$c = \frac{Q}{m\Delta T}$$

For a substance to change phase, an additional heat transfer is required. L_f is the amount of heat transferred per unit mass as the substance melts or freezes, and L_v is the heat transferred per unit mass as it vaporizes or condenses.

$$Q = \pm\, mL_f \qquad \text{or} \qquad Q = \pm\, mL_v$$

2. Identify the given information:

volume of water $V_w =$

initial temperature of water $T_{1w} =$

mass of ice $m_{ice} =$

initial temperature of ice $T_{1\text{-ice}} =$

(continued on next page . . .)

The constant values required for this problem are

specific heat of water $c_w = 4186$ J/kg C°

heat of freezing for water/ice $L_{f\text{-ice}} = 333$ kJ/kg

melting point of ice $T_{mp\text{-ice}} = 0°C$

The density of water is taken to be 1 g/cm³.

Solving the problem

3. We must pay attention to the units in the given information and convert to consistent units throughout the problem. Determine the mass, in kilograms, of the water, using 1.00 ml = 1.00 cm³.

4. In this problem, the net heat exchange will equal zero and the system will reach thermal equilibrium. The water will lose thermal energy, and the ice will use thermal energy to undergo a change of phase. If additional thermal energy is available after all the ice is melted, the ice water will be warmed. Calculate the thermal energy that would be released by lowering the water to the freezing point 0° C.

5. Determine the thermal energy required to melt all of the ice.

6. Solve for the final temperature in this case by applying the law of conservation of energy.

7. If the available energy does not equal or exceed the required energy, not all of the ice will melt. The final temperature will be 0° C. Calculate the amount of the ice that can be melted.

The Ideal Gas

Problem Description
An ideal gas is held in a container at constant volume. Initially, its temperature is 10.0° C and its pressure is 2.50 atm. What is its pressure when its temperature is 80.0° C?

Before we begin...

1. State the ideal gas law.

2. Identify the given information:

   initial temperature $T_1 =$

   final temperature $T_2 =$

   initial pressure $P_1 =$

Solving the problem

3. In solving problems using the ideal gas law, the temperature must be expressed in terms of absolute (Kelvin) temperature. Convert the temperatures T_1 and T_2 to kelvins.

4. The ideal gas law can be rewritten as

$$\frac{P}{T} = \frac{Nk_B}{V}$$

Since neither N (the number of gas molecules) nor V (the volume of the gas) change, the quantity P/T must also remain constant in this problem. Use this ratio to solve for the new pressure P_2.

The First Law of Thermodynamics

Problem Description

During a controlled expansion, the pressure of a certain gas is $P = P_0 e^{-bV}$, where $P_0 = 12$ atm, and $b = (1/12)$ m^3. Determine the work performed by the gas as it expands from 12 m^3 to 36 m^3.

Before we begin...

1. Write the expression for the work done by a gas in expanding through a volume.

2. What are the initial and final volumes of this gas?

   $V_1 =$ $V_2 =$

Solving the problem

3. Using the definition of work, evaluate the expression using the given limits of integration.

4. Make certain that the units used are appropriate for the problem.

Carnot Engines

Problem Description

A given Carnot engine has a power output of 150 kW. The engine operates between two reservoirs at 20° C and 500° C. How much thermal energy is absorbed per hour? How much thermal energy is lost per hour?

Before we begin...

1. Identify the given information:

   temperature of the cold reservoir $T_c =$

   temperature of the hot reservoir $T_h =$

   power output $P =$

2. What is the definition of power P?

3. What does the efficiency of a heat engine measure? Express your answer in terms of work done and energy required.

4. What is the equation for the efficiency of a Carnot engine in terms of the temperatures of the hot and cold reservoirs?

Solving the problem

5. The efficiency of a Carnot engine can be expressed as a function of its absolute temperatures T_c and T_h. Convert the temperatures to kelvins and solve for the efficiency of the engine.

(continued on next page . . .)

Problems

6. Use the definition of power and the given information to find the rate at which work is being done by the engine.

7. Use the definition of efficiency and the rate at which work is done to calculate the thermal energy that is absorbed per hour.

8. Using the definition of work done by a heat engine, compute Q_c, the energy that is lost to the cold reservoir.

Carnot Engines—The Heat Pump

Problem Description

What is the maximum coefficient of performance (COP) of a heat pump that brings heat from outdoors at −3° C into a 22° C house?

Before we begin...

1. The maximum coefficient of performance for a heat pump is its Carnot COP. How is this computed?

2. Identify the given information:

   temperature of hot reservoir $T_h =$

   temperature of cold reservoir $T_c =$

Solving the problem

3. The heat pump does work W, which is also available to heat the house. This is already factored into the Carnot COP. Convert the temperatures T_h and T_c to the Kelvin scale.

4. Evaluate the Carnot COP.

Problems

Applications of Entropy—General Process for an Ideal Gas

Problem Description

An airtight freezer contains air at 25°C and 1.0 atm. The air is then cooled to −18 °C. What is the change in entropy of the air if the volume is held constant? What would the change be if the pressure were maintained at 1 atm during the cooling? You can assume the air is an ideal gas, with specific heat $C_v = (5/2)R$.

Before we begin...

The processes described above involve a change in temperature and, in the second case, a change in volume. According to the first law of thermodynamics,

$$dQ_r = dU + PdV$$

where the subscripted r indicates a reversible process.

For an ideal gas, $dU = nC_VdT$, so we can write

$$dQ_r = nC_VdT + nRT(dV)/V$$

To evaluate this expression, divide both sides of the equation by the temperature T. This gives us the relationship

$$\frac{dQ_r}{T} = nC_V\frac{dT}{T} + nR\frac{dV}{V}$$

If C_V is assumed constant, integration shows us that

$$\Delta S = \int_i^f \frac{dQ_r}{T} = nC_V \ln \frac{T_f}{T_i} + nR \ln \frac{V_f}{V_i}$$

where ΔS is the change in entropy.

1. Identify the given information:

   initial temperature in kelvins $T_i =$

   final temperature in kelvins $T_f =$

   initial pressure $P_i =$

Solving the problem

2. In the problem's first question, in which volume is constant, $V_i = V_f$. Use this information to compute the change in entropy ΔS.

3. In the second situation described in the problem, the volume changes under constant pressure. Using the ideal gas law, write an expression for V in terms of the other variables. Use this information to help evaluate the change in entropy ΔS.

Applications of Entropy

Problem Description

A 100,000-kg iceberg with a temperature of −5° C breaks away from the polar ice shelf and floats into the ocean, which has a temperature of 5° C, where it completely melts. What is the change in the entropy of the world due to this process?

Before we begin...

You will need to consider the changes in entropy of both the iceberg and the surrounding ocean.

1. The key to this problem is that heat flows from the warmer ocean to the colder iceberg. How is heat flow related to the change in a system's entropy?

2. The iceberg merges with the ocean around it in a three-stage process. Identify the three stages, and begin to think about how much heat is transferred during each stage.

3. Identify the given information:

    mass of the iceberg $m =$

    initial temperature of the iceberg $T_i =$

    temperature of the ocean $T_f =$

Other information that you may find useful:

    melting point of ice $T_{\text{m-p}} = 0.0°\ \text{C} = 273.0\ \text{K}$

    latent heat of freezing for ice $L_f = 3.33 \times 10^5\ \text{J/kg}$

    specific heat of ice $c_{\text{ice}} = 2090\ \text{J/kg K}$

    specific heat of water $c_{\text{water}} = 4186\ \text{J/kg K}$

Solving the problem

4. Write an expression for the amount of heat transferred as ice warms by a small temperature increment dT. Use this expression to find the changes in entropy of the ice as it warms from $T_i = -5°$ C to $T_{m-p} = 0°$ C.

5. The ice melts at constant temperature T_{m-p}. How much heat is transferred during this process? How does the entropy of the iceberg change as it melts?

6. Find the change of entropy of the recently melted iceberg as it warms from $T_{m-p} = 0°$ C to $T_f = 5°$ C.

7. Find the change in entropy of the surrounding seawater as the above three processes take place.

8. What is the total change in entropy of the iceberg and the surrounding seawater?

MODULE 11

The Electric Field

Electric forces act over a distance between charged objects and can be either attractive or repulsive. It is convenient to think of these forces as being exerted by a vector quantity called the electric field, which can have a different magnitude and direction at each point in space. A particle's charge measures its tendency to affect and be affected by electric fields.

In this module, we examine electric fields and the related concepts of electric potential energy and electric potential. Gauss's law provides a useful method of computing the electric field in regions of high symmetry.

DEFINITIONS

Electric charge. An intrinsic feature of every atom, which can be either negative or positive. The SI unit of charge is called the coulomb.

Insulator. An electric insulator resists the movement of charge.

Conductor. An electric conductor allows charge to move easily within it.

Coulomb's law. The force between any two stationary charges is directly proportional to the product of the charges and inversely proportional to the square of the distance separating the charges. Opposite signed charges attract each other while like signed charges repel.

Electric field lines. Lines that point in the same direction as the electric force on a positive charge at every point in space. Field lines begin on positive charges and end on negative charges. No two electric field lines can cross or touch.

Electric flux, ϕ. The number of electric field lines that pass through a given area of a surface A. If a closed surface encloses no net charge, the total electric flux through the surface is zero.

Gauss's law. The total electric flux through a closed surface is proportional to the charge enclosed by that surface.

Electric potential energy. The work done when a charged particle moves through an electric field.

Electric potential. The change in electric potential energy experienced by a charge when it moves through an electric field. The SI unit is the volt.

USEFUL EQUATIONS

Coulomb's law can be expressed in the equation

$$\vec{F}_{21} = k_e \frac{q_1 q_2}{r^2} \hat{r}_{12}$$

where k_e is the Coulomb constant, q is the charge on a particle, and r is the distance separating the charges. The force is attractive if the charges are of opposite sign. It is repulsive if the charges have the same sign.

The electric field $\vec{E}$ can be defined by the relationship

$$\vec{E} = \vec{F}/q_0$$

with q_0 representing the charge experiencing the force at the particular location.

The electric potential energy is related to the electric field by

$$\Delta U = -q_0 \int_A^B \vec{E} \cdot d\vec{s}$$

The electric potential difference is computed by dividing the electric potential energy difference by the charge experiencing the field

$$\Delta V = -\int_A^B \vec{E} \cdot d\vec{s}$$

We can compute the components of the electric field by using the equations

$$E_x = -\frac{\partial V}{\partial x} \qquad E_y = -\frac{\partial V}{\partial y} \qquad E_z = -\frac{\partial V}{\partial z}$$

Gauss's law is expressed by the equation

$$q_{net} = \epsilon_0 \oint \vec{E} \cdot d\vec{A}$$

where ϵ_0 is a constant called the permittivity of free space.

Coulomb's Law

Problem Description

Four identical point charges of $+10.0\ \mu C$ each are located on the corners of a rectangle as shown. The dimensions of the rectangle are $L = 60.0$ cm and $W = 15.0$ cm. Calculate the magnitude and direction of the net electric force exerted on the charge at the lower-left corner by the other three charges.

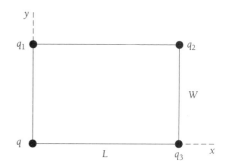

Before we begin...

1. Identify the given information (express length and width in meters):

     charge on each particle $q =$

     length of rectangle $L =$

     width of rectangle $W =$

2. State Coulomb's law for the force between two point charges.

3. How is the direction of the force between two point charges determined?

4. Draw the free-body diagram for the charge whose net force we are to calculate. Use the upper left charge as q_1, upper right q_2, and lower right q_3.

Solving the problem

5. The electric force $\vec{F}_T$ is equal to the sum of all of the electric forces $\vec{F}_i$ that act upon it. Using Coulomb's law, evaluate the forces $\vec{F}_1$, $\vec{F}_2$, and $\vec{F}_3$.

6. Resolve the forces into their x and y components. Add the x components to yield the total x force. Do the same for the y components.

7. Use the Pythagorean theorem to determine the magnitude of $\vec{F}_T$.

8. Determine the direction of $\vec{F}_T$ by applying the definition of the trigonometry function $\tan \theta$.

The Electric Field and Field Lines

Problem Description

Four point charges are at the corners of a square of side a as shown. Determine the magnitude and direction of the electric field at the location of charge q. What is the resultant force on q?

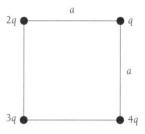

Before we begin...

1. Assume that the charge q is located at the origin of a cartesian coordinate system. Sketch the electric fields at the origin due to the other charges.

2. How is the electric field for a point charge calculated?

3. How is the direction of the electric field due to a point charge determined?

4. What is the electric force exerted on a charge q by a known field $\vec{E}$?

Solving the problem

5. The electric field $\vec{E}_T$ is equal to the sum of all of the electric fields $\vec{E}_i$ that act upon it. Using the relationship for the electric field due to a point charge, evaluate the fields $\vec{E}_{2q}$, $\vec{E}_{3q}$, and $\vec{E}_{4q}$.

6. Resolve the fields into their x and y components. Add the x components to yield the total x field. Do the same for the y components.

7. Use the Pythagorean theorem to determine the magnitude of $\vec{E}_T$.

8. The direction can be determined by applying the definition of the trigonometry function tan θ.

9. Using the definition of electric fields, find the electric force on charge q by applying $\vec{F} = q\vec{E}$.

Gauss's Law—Electric Flux

Problem Description
A uniform electric field $a\hat{i} + b\hat{j}$ intersects a surface of area A. What is the flux through this area if the surface lies in the *yz* plane? What is the flux through this area if the surface lies in the *xz* plane? What is the flux through this area if the surface lies in the *xy* plane?

Before we begin...	The surface vector $\vec{A}$ is drawn perpendicular to the surface A.

1. How is electric flux calculated?

2. Identify the given information:

   electric field $\vec{E} =$

Solving the problem

3. Assuming that the magnitude of the surface vector is A, express the surface vectors $\vec{A}_1$, $\vec{A}_2$, and $\vec{A}_3$ that correspond to the three surfaces described.

4. Evaluate the flux through each surface by using the definition of electric flux.

Gauss's Law

Problem Description
The total electric flux through a closed surface in the shape of a cylinder is 8.60×10^4 N·m²/C. What is the net charge within the cylinder? How would your answers change if the net flux were -8.60×10^4 N·m²/C?

Before we begin...

1. State Gauss's law for electric fields.

2. Because the total flux is known, does the shape of the closed surface matter when calculating the net charge within the surface? Explain.

3. If the electric flux is negative, what does this tell us about the direction of $\vec{E}$ as related to $\vec{A}$?

Solving the problem

4. Use the relationship between total electric flux and the net charge to calculate the charge.

5. Using the properties of the scalar (dot) product, determine the sign of the charge distribution.

Examples of the Electric Field

Problem Description
The charge per unit length on a long, straight filament is −90.0 μC/m. Find the electric field 10.0 cm, 20.0 cm, and 100 cm from the filament, where distances are measured perpendicular to the length of the filament.

Before we begin...

1. Assume that the charge per unit length is represented by λ. How much charge would be inside a right cylinder of length l and radius r as shown in the diagram?

2. The electric field $\vec{E}$ by symmetry will point radially inward toward the wire, and therefore it will be antiparallel to $d\vec{A}$ through the side of the cylinder. What will the flux be through the two end caps of the cylinder?

Solving the problem

3. Apply Gauss's law to the drawing to derive the electric field at radial distances from a long straight wire.

4. Use this result to evaluate the magnitude of $\vec{E}$ at the given distances.

Electric Potential

Problem Description

Two point charges, $q_1 = +5.00$ nC and $q_2 = -3.00$ nC, are separated by 35.0 cm. What is the potential energy of the pair? What is the significance of the algebraic sign of the answer? What is the electric potential at a point midway between the charges?

Problems

Before we begin...

The change in potential energy experienced when a charge q_0 is moved from a point A to a point B in the presence of an external electric field $\vec{E}$ is computed by

$$\Delta U = -q_0 \int_A^B \vec{E} \cdot d\vec{S}$$

If the field is due to a second point charge, we have

$$\vec{E} = k_e \frac{q}{r^2} \hat{r}$$

By setting $U = 0$ at $r = \infty$, the potential energy at position r can be calculated:

$$U = -q_0 \int_\infty^r k_e \frac{q}{r^2} dr = k_e \frac{q q_0}{r}$$

1. How is electric potential difference related to the change in electric potential energy?

(continued on next page . . .)

2. Identify the given information:

   charge on first particle $q_1 =$

   charge on second particle $q_2 =$

   distance between the particles $r =$

Solving the problem

3. Use the given information to compute the potential energy. If the potential energy is negative, work will have to be done to separate the two charges.

4. Use the definition of electric potential to calculate the voltage from the potential energy.

The Electric Field and Electric Potential

Problem Description
The electric potential in a certain region is
$V = (4xz - 5y + 3z^2)$ volts. Find the magnitude of the electric field at $(2, -1, 3)$, where all distances are in meters.

Before we begin...   1. How is the electric field related to the electric potential?

2. Identify the given information:

electric potential $V =$

position at which to evaluate field $\vec{r} = (x, y, z) =$

Solving the problem   3. Compute the components of $\vec{E}$ using the relationship between $\vec{E}$ and V.

4. Evaluate the components of $\vec{E}$ using the given position in space.

Problems

The Electric Field and Electric Potential

Problem Description

When an uncharged conducting sphere of radius a is placed at the origin of an xyz coordinate system that lies in an initially uniform electric field $\vec{E} = E_0\hat{k}$, the resulting electrostatic potential is $V(x, y, z) = V_0$ for points inside the sphere, and

$$V(x, y, z) = V_0 - E_0 z + \frac{E_0 a^3 z}{(x^2 + y^2 + z^2)^{3/2}}$$

for points outside the sphere, where V_0 is the (constant) electrostatic potential on the conductor. Use this equation to determine the x, y, and z components of the resulting electric field.

Before we begin...

1. In this problem, the potential has a different form within the sphere than it does outside the sphere. If the electric potential is constant over a region of space, what does this tell us about the electric field over that same region of space?

Solving the problem

2. For the region inside the sphere, use the answer to the question above to evaluate the electric field.

3. Find the functional relationship between the components of the electric field and the electric potential outside the sphere.

MODULE 12 The Magnetic Field

In Module 11, *The Electric Field*, we discussed the interaction between charged objects in terms of electric fields. In this module, we find that moving charges have another interaction, which we describe in terms of the magnetic field. Magnetic and electric forces are both present when a charged particle moves through a region of space containing magnetic and electric fields. The magnetic field does no work on a moving charged particle because the force exerted is at right angles to the path taken by the particle.

We will explore the links between electric and magnetic fields and discover that changing magnetic fields produce electric fields, and vice versa.

DEFINITIONS

Tesla. The SI unit for the magnetic field $\vec{B}$. One tesla T equals one newton per ampere·meter.

Magnetic force. The force exerted on a moving charged particle by an external magnetic field is

$$\vec{F}_B = q(\vec{v} \times \vec{B})$$

Ampère's law. A law that describes a fundamental relationship between constant currents and the magnetic fields they produce.

$$\mu_0 I_{net} = \oint \vec{B} \cdot d\vec{s}$$

This equation is highly useful for situations with symmetry. The right-hand rule determines the direction of the associated magnetic field for a current carrying wire.

Gauss's law for magnetism. A law asserting that the magnetic flux through any closed surface is zero. This gives rise to the statement that magnetic monopoles do not exist.

Induction. The process by which changing magnetic flux gives rise to a potential difference (emf). This comes from Faraday's law of electromagnetic induction.

Faraday's law of induction is expressed by the equation

$$\text{Emf} = -\frac{d\Phi_B}{dt}$$

Magnetic flux is computed by

$$\Phi_B = \oint \vec{B} \cdot d\vec{A}$$

The magnetic flux can be changed by changing either the magnetic field, the area through which the field points, or the angle between the field and the surface normal.

Lenz's contribution to Faraday's law is the negative sign. The induced emf opposes the change in magnetic flux.

USEFUL EQUATIONS

The Lorentz force acting upon a moving charged particle in a region of space with electric and magnetic fields is

$$\vec{F} = q[\vec{E} + (\vec{v} \times \vec{B})]$$

The force on a long straight current carrying wire is

$$F = ILB \sin \theta$$

where θ is the angle between L and B.

Ampère's law states

$$\mu_0 I_{net} = \oint \vec{B} \cdot d\vec{s}$$

where μ_0 is a constant called the permeability of free space. This assumes that the magnetic field is in free space.

The general form of Faraday's law allows for induced electric fields and is given by

$$\oint \vec{E} \cdot d\vec{s} = -\frac{d\Phi_B}{dt}$$

The induced electric field as a result of a changing magnetic flux is not conservative.

Saunders Core Concepts in Physics Workbook

The general form of Ampère's law involves a displacement current I_d in addition to the conduction current I. The displacement current is defined by the equation

$$I_d = \epsilon_0 \frac{d\Phi_E}{dt}$$

Using the displacement current, we express Ampère's law as

$$\oint \vec{B} \cdot d\vec{s} = \mu_0 (I + I_d)$$

Magnetic Force on a Moving Charge

Problem Description

The magnetic field of the Earth in a certain region is uniform, directed vertically downward, with a magnitude of 0.5×10^{-4} T. A proton is moving horizontally toward the west in this field with a speed of 6.2×10^6 m/s. What are the direction and magnitude of the magnetic force the field exerts on this charge? What is the radius of the circular arc followed by this proton?

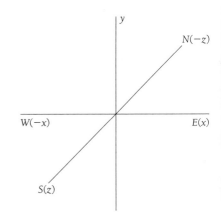

Before we begin...

1. How is magnetic force $\vec{F}_B$ related to the magnetic field $\vec{B}$ and the velocity v of a moving charged particle?

2. Does this force change the speed of the particle? If not, what type of acceleration does the particle experience?

3. Draw, not to scale, the magnetic field and velocity on the coordinate axes shown above.

4. Identify the given information:

   magnetic field $\vec{B}$ =

   proton's velocity $\vec{v}$ =

   proton's charge q =

   proton's mass m =

Solving the problem

5. Substitute the given information into the relationship between $\vec{F}_B$, $\vec{v}$, and $\vec{B}$, and solve for the force $\vec{F}_B$.

6. How is centripetal force related to mass m, speed v, and the radius r of the path?

7. Substitute the given and calculated information into the centripetal force equation to compute the radius of the path.

The Lorentz Force

Problem Description
An electron moves through a uniform electric field $\vec{E} = (2.5\hat{i} + 5.0\hat{j})$ V/m and a uniform magnetic field $\vec{B} = 0.40\hat{k}$ T. Determine the acceleration of the electron when it has a velocity $\vec{v} = 10\hat{i}$ m/s.

Before we begin...

1. How does the Lorentz equation relate force $\vec{F}$, electric field $\vec{E}$, magnetic field $\vec{B}$, and charge q?

2. How is the acceleration of a particle computed if the net force $\vec{F}$ and the mass m are known?

3. Identify the given information:

   electric field $\vec{E} =$

   magnetic field $\vec{B} =$

   electron's velocity $\vec{v} =$

   electron's charge $e =$

   electron's mass $m_e =$

Solving the problem

4. The net force acting upon the charged particle is equal to the sum of its electrical force $\vec{F}_E$ and its magnetic force $\vec{F}_B$. Evaluate the electric force using $\vec{F}_E = q\vec{E}$.

5. Evaluate the magnetic force $\vec{F}_B$.

6. Set the sum of the forces equal to the mass multiplied by the acceleration of the electron and solve for the acceleration.

Ampère's Law

Bundle of 100 wires
carrying current into page

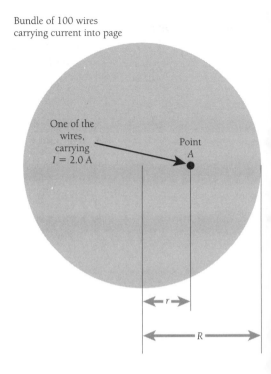

One of the wires, carrying $I = 2.0$ A

Point A

Problem Description

A packed bundle of 100 long, straight, in-sulated wires forms a cylinder of radius $R = 0.50$ cm. Each of the 100 wires carries a current of 2.0 amps "into the page" as shown. What are the magnitude and di-rection of the magnetic force per unit length on one of the wires that passes through point A on the diagram, at a distance $r = 0.20$ cm from the center of the bundle?

Before we begin...

1. What is the expression for the magnetic force per unit length on a long, straight wire carrying a current $\vec{I}$ in a magnetic field $\vec{B}$?

2. Because the bundle is circularly symmetric about its center, Ampère's law should let us find the magnetic field $\vec{B}$ at point A. Draw an amperian loop that passes through point A and takes advantage of the bundle's symmetry.

3. What is the total current I_{in} passing through the amperian loop you have drawn?

4. How does Ampère's law relate the line integral of the magnetic field around your amperian loop to the current I_{in} passing through it?

Solving the problem

5. Use Ampère's law to find the strength of the magnetic field at point A. Which direction does this field point?

6. What are the magnitude and direction of the force per unit length exerted by this field on the wire passing through point A?

Magnetic Flux and Gauss's Law for Magnetism

Problem Description

Consider the cube having a side length L as shown in the figure. A uniform magnetic field $\vec{B}$ is directed perpendicular to face *abfe*. Find the magnetic flux through the imaginary planar loops *dfhd* and *acfa*.

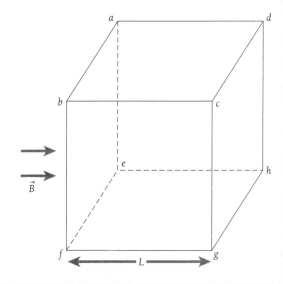

Before we begin...

1. How is magnetic flux calculated?

2. Because $\vec{B}$ is a constant and the surfaces are flat, we need to find the projection of the surfaces onto a surface that has a normal vector parallel to $\vec{B}$—i.e., *abfe* or *dcgh*. The projection of a surface onto another surface is like a surface casting a shadow onto the other. This projection is calculated by $A \cos \theta$, where θ is the angle between the two surfaces. Sketch the imaginary planar loop *dfhd* and its projection onto *dcgh*.

Solving the problem

3. Project the surface *dfhd* onto the surface *dcgh*. If the area of *dcgh* is L^2, what is the area of the projection of *dfhd* onto *dcgh*?

   Use this information to compute the magnetic flux through *dfhd*.

4. Repeat the process, projecting *acfa* onto *abfe*.

Faraday's Law of Induction and Lenz's Law

Problem Description

A flat loop of wire of area 14 cm² and two turns is perpendicular to a magnetic field whose magnitude decays in time according to $B = B_0 e^{-t/t_0}$ where $B_0 = 0.50$ T and $t_0 = 7$ s. What is the induced emf as a function of time?

Before we begin...

1. State Faraday's law of induction.

2. How is the total flux calculated at any given instant?

3. Identify the given information:

   area of loop $A =$

   number of turns $N =$

   magnetic field $B =$

   angle between $\vec{B}$ and $\vec{A}$ $\theta =$

Solving the problem

4. Because we are not concerned with the sign of the induced emf, Lenz's contribution to Faraday's law will not be used in this problem. Using the given information, write the expression for the magnetic flux as a function of time.

5. Apply Faraday's law to the changing flux to evaluate the emf.

Faraday's Law of Induction and Lenz's Law

Problem Description

A metal bar spins at a constant rate about one fixed end in the magnetic field of the Earth. The rotation occurs in a region where the component of the Earth's magnetic field perpendicular to the plane of rotation is 3.3×10^{-5} T. If the bar is 1.0 m in length and its angular speed is 5π rad/s, what potential difference is developed between its ends?

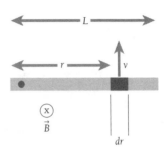

Before we begin...   Computing the emf induced in a rotating bar we assume that the uniform magnetic field $\vec{B}$ is pointing into the plane of this page. The bar, pivoted about one end, is rotating with an angular speed $\omega = v/r$.

Each element of the bar, dr, is moving perpendicular to the field and produces an emf

$$d\varepsilon = Bvdr = B\omega rdr$$

To find the total emf ε, we must integrate over the entire bar segments.

$$\varepsilon = \int_0^L B\omega rdr = B\omega\frac{L^2}{2}$$

1. Identify the given information:

   magnetic field (component into page) $B =$

   length of bar $L =$

   angular speed of bar $\omega =$

Solving the problem   2. Substitute into the equation for the emf of a rotating bar and evaluate the emf.

General Form of Faraday's Law

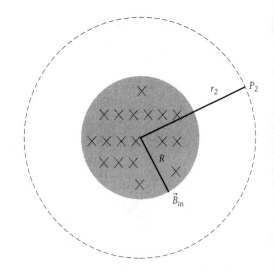

Problem Description

For the situation described in the figure, $r_2 = 2R = 5.0$ cm, and the magnetic field changes with time according to $B = (2.0\ t^3 - 4.0t^2 + 0.80)$ T, where t is in seconds. Calculate the magnitude and direction of the force exerted on an electron located at point P_2 when $t = 2.0$ s. At what time is this force equal to zero?

Before we begin...

1. The force that the electron will experience in this problem is due to the induced electric field $\vec{F}$ that arises as a result of the changing magnetic flux with respect to time. State the general form of Faraday's law.

2. How is the force acting upon a charged particle related to the electric field creating the force?

3. Identify the given information:

   magnetic field in central region $B =$

   radius of central region $R =$

   electron's distance from center $r_2 =$

4. Write the expression for the magnetic flux through the surface bounded by the dotted line.

5. Take the derivative of the magnetic flux with respect to time to evaluate the emf as a function of time.

6. If we draw a circle of radius r_2 that passes through the point P_2, we know that the electric field will be constant along the circle, so that

$$\int \vec{E} \bullet d\vec{S} = E(2\pi r_2)$$

Set this expression equal to the rate of change of magnetic flux.

7. Solve for E and evaluate at the given time.

8. Solve for the force that results from the electric field.

Because the electric field will be directed along the same path as the induced emf, the direction of the force on the electron can be deduced. What is the direction of the induced emf?

Determine the direction of the force on the electron.

9. The force will be equal to zero when the electric field equals zero. Calculate the time at which this occurs.

Electric Circuits

In this module, we discuss simple electric currents. We examine three devices: resistors, capacitors, and inductors. Resistors are nonconservative devices that convert electric energy into thermal energy. Capacitors and inductors are conservative devices, that store energy in their electric and magnetic fields respectively. The laws of conservation of charge and energy can be employed to analyze circuits containing these devices.

DEFINITIONS

Electric current, I. The rate at which electric charge flows past any point in a conductor. The SI unit for current is the ampere, which is one coulomb per second.

Resistor. A device that resists current flow when a potential difference is applied across the device. The resistance R of an object depends on its size, shape, and temperature, as well as the material of which it is made. The SI unit is the ohm.

Series connection. A connection that occurs when there are no junction points between elements.

Parallel connection. A connection that occurs when the path of a current divides elements.

Capacitor. A device that stores energy within its electric field. The capacitance C depends upon the geometry and materials of the capacitor. The SI unit of capacitance is the farad, which is one coulomb per volt.

Inductor. A device that stores energy within its magnetic field. The inductance L depends upon the geometry and materials of the inductance coils. The SI unit of inductance is the henry, which is one volt·second per ampere.

Ohm's law relates the voltage V to the electric current I and the resistance R by the equation

$$V = IR$$

Resistors connected in series have an equivalent resistance computed by

$$R_S = R_1 + R_2 + \ldots$$

Resistors connected in parallel have an equivalent resistance computed by

$$\frac{1}{R_p} = \frac{1}{R_1} + \frac{1}{R_2} + \ldots$$

Kirchhoff's laws apply the conservation of charge to the junction rule:

$$\sum I_{\text{in}} = \sum I_{\text{out}}$$

and the conservation of energy to the loop rule:

$$\sum \Delta V = 0 \text{ around the closed loop}$$

Capacitance is related to voltage and charge by the equation

$$Q = CV$$

The energy stored in a capacitor is computed by the relationship

$$U = \frac{1}{2}QV = \frac{1}{2}CV^2 = \frac{Q^2}{2C}$$

Capacitors in series have an equivalent capacitance found by

$$\frac{1}{C_S} = \frac{1}{C_1} + \frac{1}{C_2} + \ldots$$

Capacitors in parallel have an equivalent capacitance computed as

$$C_p = C_1 + C_2 + \ldots$$

Inductance is related to the induced voltage opposing the change in current and the rate of change of electric current by the equation

$$V = L\frac{dI}{dt}$$

The energy stored in an inductor is computed by the relationship

$$U = \frac{1}{2}LI^2$$

Inductors in series and parallel have equivalent inductances computed in the same manner as resistors in series and parallel.

An LC circuit oscillates with an angular frequency given by

$$\omega_0 = \frac{1}{\sqrt{LC}}$$

An RLC circuit oscillates with an angular frequency expressed by

$$\omega = \sqrt{(\omega_0^2 - \beta^2)}$$

where ω_0 is the natural frequency of oscillation for an *LC* circuit and $\beta = R/2L$.

Voltage, Resistance, and Ohm's Law

Problem Description
Batteries are rated in terms of ampere hours (A·h), where a battery rated at 1.0 A·h can produce a current of 1.0 A for 1.0 h. What is the total energy in kilowatt hours, stored in a 12.0 V battery rated at 55.0 A·h? At $0.06 per kilowatt hour, what is the value of the electricity produced by this battery?

Problems

Before we begin...

1. From the principles of work and energy, we recall that power is defined as the rate at which work is done or energy is consumed. How is power computed for electric circuits?

   The total energy can be computed by multiplying the power by the time.

2. Identify the given information:

   voltage $V =$

   cost/kW·h $=$

   rating $=$

(continued on next page . . .)

Solving the problem

3. A battery rated at 55.0 A·h can deliver a current of 55.0 A for a time of 1.0 hours. It can also deliver 5.0 A for 11.0 h, or any other product of current and time equaling 55.0 A·h. Write the expression for the energy delivered in terms of current, voltage, and time.

4. What are the dimensions of the equation?

5. By what would we have to multiply the rating of the battery to give the dimensions in the above equation?

6. Use the information given to solve for the energy stored in the battery.

7. Compute the cost of the total energy by multiplying the cost per kW·h by the number of kW·h stored in the battery.

Circuit Analysis and Kirchhoff's Laws

Problem Description

The resistance between terminals *a* and *b* in the figure is 75 Ω. If the resistors labeled *R* have the same value, determine *R*.

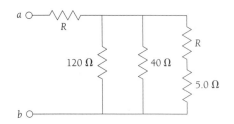

Before we begin...

1. Many times we find that a connection of elements can be simplified by identifying the elements in simple series and in simple parallel. What is the equivalent resistance for resistors connected in simple series?

2. What is the equivalent resistance for resistors connected in simple parallel?

Solving the problem

3. As we attempt to simplify the connection of resistors, observe the two resistors on the far right side of the diagram. How are they connected to each other?

4. Compute and write their equivalent resistance as R_1.

5. Redraw the diagram using R_1.

(continued on next page . . .)

6. In the redrawn diagram, how are R_1, the 120 Ω resistor, and the 40 Ω resistor connected?

Compute their equivalent resistance as R_2.

7. Redraw the diagram using R_2.

8. In the redrawn diagram, how are R_2 and R connected?

Write the equation for their equivalent resistance R_E.

9. Because $R_E = 75$ Ω, substitute the value of R_2 and subsequently R_1 into the equation and solve for the value of R.

Circuit Analysis and Kirchhoff's Laws

Problem Description
Find the potential difference between points *a* and *b* in the diagram. Find the currents I_1, I_2, and I_3 as shown.

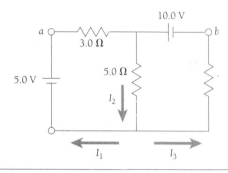

Before we begin...

1. This problem has multiple sources of emf and multiple loops. Solving it requires Kirchhoff's laws. State the junction rule and the loop rule.

Solving the problem

2. Select a junction point in the diagram and apply the junction rule.

There are three unknown currents in this problem, so we need three independent equations to simultaneously solve for them. The junction rule gives us one such equation. The loop rule will give us the other two.

3. Select a loop in the diagram and apply the loop rule. Repeat the process for another loop.

4. Solve the equations for the values of the currents I_1, I_2, and I_3.

5. To evaluate the potential difference between *a* and *b*, travel any path between the two and use the algebraic sum of the potentials.

Capacitors

Problem Description
For the system of capacitors shown in the diagram, find the equivalent capacitance of the system, the potential across each capacitor, the charge on each capacitor, and the total energy stored by the group.

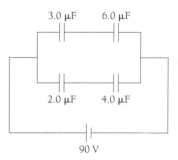

Before we begin...

1. Many times a circuit can be simplified by identifying elements in simple series and in simple parallel. What is the equivalent capacitance for capacitors in series?

2. What is the equivalent capacitance for capacitors in parallel?

3. What do capacitors in series have in common?

   What do capacitors in parallel have in common?

4. How is the energy stored in a capacitor computed?

Solving the problem

5. How are the 3.0 μF capacitor and the 6.0 μF capacitor connected?

   Write the equivalent capacitance of the two as C_2.

6. How are the 2.0 μF capacitor and the 4.0 μF capacitor connected?

   Write the equivalent capacitance of the two as C_3.

7. Redraw the diagram using C_2 and C_3. How are C_2 and C_3 connected?

   Evaluate the equivalent capacitance C_T.

8. Use the answers to the above questions to help evaluate the charge and potential difference across each capacitor.

9. Using the equivalent capacitance, evaluate the energy stored by this capacitor. This is the same as the sum of the energies stored in all of the combined capacitors.

Inductors

Problem Description

Consider the circuit shown in the diagram. What energy is stored in the inductor when the current reaches its final equilibrium value after the switch S is closed?

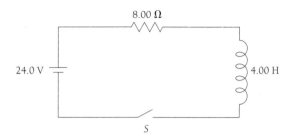

Before we begin...

1. An inductor sets up a potential difference to oppose changes in the current flowing in a circuit. How is this potential difference expressed in terms of the inductance L?

2. Once the circuit has reached equilibrium, what is the rate of change of the current through the inductor?

3. How is the energy stored in the magnetic field of an inductor computed?

4. Identify the given information:

    battery voltage $V =$

    resistance $R =$

    inductance $L =$

Solving the problem

5. Once the opposing potential difference has been reduced to zero, the current flowing through the circuit reaches its maximum value. Use Ohm's law to compute the value of the maximum current.

6. Use the maximum value of the current I and the given inductance L to calculate the energy stored in the magnetic field of the inductor.

Circuits Containing Resistors, Inductors, and Capacitors

Problem Description

Calculate the inductance of an *LC* circuit that oscillates at a frequency $f = 120$ Hz when the capacitance is 8.00 μF.

Before we begin...

An *LC* oscillating circuit is the electrical equivalent of the mechanical spring-mass system oscillating without friction. From Kirchhoff's loop rule, we obtain

$$L\frac{d^2Q}{dt^2} + \frac{Q}{C} = 0$$

which is analgous to

$$m\frac{d^2x}{dt^2} + kx = 0$$

The solution to the differential equation has already shown that the system will oscillate with an angular frequency

$$\omega_0 = \sqrt{k/m}$$

For the *LC* circuit, direct comparison allows us to predict

$$\omega_0 = \frac{1}{\sqrt{LC}}$$

1. How is the angular frequency ω_0 related to the frequency f?

2. Identify the given information:

    frequency $f =$

    capacitance $C =$

(continued on next page . . .)

Solving the problem

3. Compute the angular frequency ω_0 for this problem.

4. Solve the equation relating the angular frequency to the inductance and capacitance to find the value of the inductance required.

Circuits Containing Resistors, Inductors, and Capacitors

Problem Description

Consider a series *LC* circuit in which *L* = 2.18 H and *C* = 6.00 nF. What is the maximum value of a resistor that, inserted in series with *L* and *C*, allows the circuit to continue to oscillate?

Problems

Before we begin...

The description of this problem is that of a damped circuit that has an angular frequency

$$\omega = \sqrt{(\omega_0{}^2 - \beta^2)}$$

where ω_0 is the natural frequency of oscillation for an *LC* circuit and $\beta = R/2L$. For small values of *R*, ω is only slightly less than ω_0.

1. What is the relationship between ω_0, *L*, and *C*?

2. Identify the given information:

   inductance *L* =

   capacitance *C* =

Solving the problem

3. Because we wish the circuit to oscillate, β must be less than ω_0. Set the equations for ω_0 and for β equal to each other and solve for the limiting value of *R*.

MODULE 14   **Geometric Optics**

Visible light consists of electromagnetic waves with wavelengths so small that noticeable diffraction does not occur as it passes through normal sized openings. In this module, we study the characteristics of light rays that point along the direction of propagation of these waves. We will examine reflection and refraction of light at a boundary between different media and the formation of images using mirrors and thin lenses.

DEFINITIONS

Reflection. When a light ray traveling in a medium reaches a boundary with a second medium, part or all of the ray is reflected back into the first medium. The angle of incidence with respect to the surface normal is equal to the angle of reflection with respect to the surface normal at the point of reflection.

Refraction. When a light ray traveling in a medium reaches a boundary with a second medium, part of the ray may be transmitted into the second medium. As a result of the speed with which light travels through the different mediums, the wavelength of the light will change in the second medium, while the frequency of the light remains unchanged.

Index of refraction. In a material n, the ratio of the speed of light as measured in a vacuum to the speed of light as measured in the material.

Critical angle, θ_C. The angle of incidence such that the angle of refraction is exactly 90°.

Lateral magnification. The ratio of the height of an image to the height of the object producing the image.

Virtual image. An image in which the light rays do not actually converge to the image point, although they appear to diverge from that point.

Focal length. Rays parallel to the principal axis of a mirror or thin lens converge on a single point known as the focal point. The distance from the focal point to the mirror or lens is called the focal length f.

USEFUL EQUATIONS

The law of reflection requires that $\theta_1 = \theta_1'$.

Snell's law for refraction is expressed as

$$n_1 \sin \theta_1 = n_2 \sin \theta_2$$

where θ_2 is the angle with respect to the surface normal that the refracted light ray makes.

The mirror equation and the thin lens equation both are given by

$$\frac{1}{f} = \frac{1}{d_o} + \frac{1}{d_i}$$

with d_o representing the object distance and d_i being the image distance. When d_i is a negative number, the image is virtual and erect; otherwise the image is real and inverted in orientation from the forming object.

The lateral magnification is computed by the equation

$$M = \frac{h_i}{h_o} = -\frac{d_i}{d_o}$$

The lens maker's equation relates the index of refraction of the lens, with respect to the surrounding medium, and the radii of curvature of the lens to the focal length of the lens.

$$\frac{1}{f} = (n - 1)\left(\frac{1}{R_1} - \frac{1}{R_2}\right)$$

Reflection

Problem Description
Determine the minimum height of a vertical flat mirror in which a person of height *h* can see his or her full image.

Before we begin...

1. State the law of reflection.

2. Draw a ray diagram illustrating the problem. Locate the eye position, the top of the head and the feet of the person.

Solving the problem

Use the law of reflection and assume that the vertical distance from the eyes to the top of the head is y_h, while the vertical distance from the eyes to the feet is y_f.

3. How far above the eyes must the mirror top be placed so that the person still can see the top of his head? Draw a ray diagram for this portion of the problem.

4. How far below the eyes must the mirror bottom be placed so that the person can still see his feet? Draw a ray diagram for this portion of the problem.

5. Add these two distances together to get the minimum height of the mirror.

Snell's Law

Problems

Problem Description
A light ray in air is incident on a water surface at an angle of 30.0° with respect to the normal to the surface. What is the angle of the refracted ray relative to this normal?

Before we begin...

1. Sketch the problem, labeling all angles.

2. State Snell's law.

3. Identify the given information:

angle of incidence $\theta_1 =$

air's index of refraction $n_1 =$

water's index of refraction $n_2 =$

Solving the problem

4. In this problem, the light is moving from air (n_1) into water (n_2). Apply Snell's law to solve for the angle at which the light will travel with respect to the normal while in the water.

Snell's Law

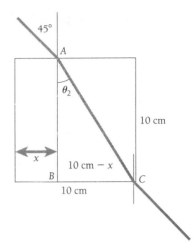

Problem Description

A glass block having $n = 1.52$ and surrounded by air measures 10.0 cm × 10.0 cm. For an angle of incidence of 45.0°, what is the maximum distance x as shown in the figure so that the ray will emerge from the opposite side?

Before we begin...   In this problem we are examining the light during the time that it is traveling through the glass. The angle that the light makes with the normal will be determined by Snell's law and the given information.

1. Identify the given information:

angle of incidence $\theta_1 =$

air's index of refraction $n_1 =$

block's index of refraction $n_2 =$

Solving the problem   2. Use Snell's law and the given information from above to calculate θ_2.

3. Examine triangle ABC in the above diagram. What is the relationship between θ_2 and the sides of the triangle, 10 cm and $(10\ cm - x)$?

4. Use this relationship to solve for x.

PROBLEM 95 (SCREEN 14.5)

Total Internal Reflection

Problem Description

A fiber optic cable (n = 1.50) is submerged in water (n = 1.33). What is the critical angle for light to stay inside the cable?

Problems

Before we begin...

1. What do we mean by *critical angle?*

2. Identify the given information:

   cable's index of refraction n_1 =

   water's index of refraction n_2 =

   angle of transmission into water θ_2 =

Solving the problem

3. Apply Snell's law and the definition of the critical angle to compute θ_C.

Flat and Spherical Mirrors

Problem Description
A concave mirror has a focal length of 40.0 cm. Determine the object position for which the resulting image is upright and four times the size of the object.

Before we begin...

1. How is lateral magnification defined? How is it related to the image and object positions?

2. Write the mirror equation.

3. Are upright images real or virtual? Is the focal length of a concave mirror positive or negative?

4. Identify the given information:

    lateral magnification $M =$

    mirror's focal length $f =$

Solving the problem

5. Use the definition of lateral magnification to give the relationship between the position of the object d_o and the position of the image d_i.

    Substitute the given information and the relationship between d_o and d_i into the mirror equation.

    Evaluate to find d_o.

Flat and Spherical Mirrors

Problem Description
An object is 15 cm from the surface of a reflective spherical Christmas tree ornament that is 6.0 cm in diameter. What are the magnification and position of the image?

Problems

Before we begin...

1. How is the focal length of a spherical mirror related to the radius of curvature of the mirror?

2. Is the focal length of a convex mirror (such as this ornament) a positive or a negative number?

3. Identify the given information:

    object distance $d_o =$

    radius of ornament $R =$

Solving the problem

4. Use the given information and the above answers to determine the focal length of the mirror.

5. Substitute the known information into the mirror equation to solve for the image distance d_i.

6. Use the equation for lateral magnification to compute M.

Thin Lenses

Problem Description

An object located 32 cm in front of a lens forms an image on a screen 8.0 cm behind the lens. Find the focal length of the lens. Determine the magnification. Is the lens converging or diverging?

Before we begin...

1. If an image is projected onto a screen, is the image real or virtual?

2. State the thin lens equation.

3. How is lateral magnification computed for a thin lens?

4. Identify the given information:

   object distance $d_o =$

   image distance $d_i =$

Solving the problem

5. Use the answers from above to determine whether the image distance d_i is positive or negative.

   Substitute into the thin lens equation to find the focal length f. If the focal length is positive, the lens is converging. If the focal length is negative, the lens is diverging.

6. Substitute the given information into the equation for lateral magnification to solve for M.

Thin Lenses

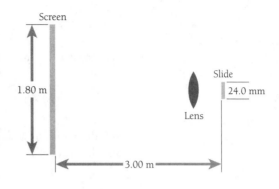

Problem Description

A slide 24.0 mm high is to be projected so that its image fills a screen 1.80 m high. The screen is 3.0 m from the slide, as shown. How far from the slide should the projector's lens be in order to form the image on the screen? What must be the focal length of the lens if the image is to be in focus?

Before we begin...

1. How is magnification defined? For a thin lens, how is magnification related to image and object distances d_i and d_o?

2. State the thin lens equation, and note the relationship it provides between d_i, d_o, and the focal length f.

Solving the problem

3. What is the magnification of the image in this problem? Use this result to write a relationship between the image and object distances, d_i and d_o.

(continued on next page . . .)

4. What other relation between d_i and d_o does the problem directly provide? (*Hint*: If you were told that $d_i = 150$ m and $d_o = 2$ m, would you consider this an accurate statement?)

5. Solve these two equations together for d_i and d_o.

6. Now that you know d_i and d_o, the thin lens equation lets you deduce the lens's focal length f.

Thin Lenses

Problem Description

A converging lens of focal length 20 cm and a diverging lens of focal length −30 cm are placed 40 cm apart, as shown. If a coin of diameter 2 cm is placed one meter to the left of the converging lens, where is the final image of the coin cast by the two-lens system? What is the diameter of this image? Is it upright or inverted?

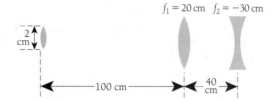

$f_1 = 20\,\text{cm}$ $f_2 = -30\,\text{cm}$

Before we begin...

In a multiple lens system, the image cast by each lens becomes the object of the next lens. The thin-lens equation applies to each individual lens.

1. How is the magnification of the image cast by each lens related to its image and object distances?

2. What is the object distance d_{o1} of the first (converging) lens?

3. What are the focal lengths f_1 and f_2 of the first and second lenses?

Solving the problem

4. Using the thin lens equation, find the position of the image cast by the first lens.

(continued on next page . . .)

5. What is the magnification of the first image? Use this to find its diameter. Is it upright or inverted?

6. What is the object distance d_{o2} of the second (diverging) lens?

7. Using the thin lens equation, find the position of the image cast by the second lens.

8. What is the magnification of the second image? Use this to find its diameter. Is it upright or inverted?

SOLUTIONS

PROBLEM 1 Coordinate Systems

Before we begin...

1.

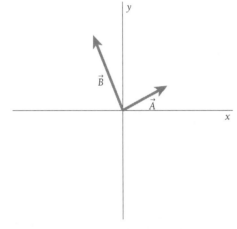

$\vec{A} = (2.50 \text{ m}, 30.0°)$

$\vec{B} = (3.80 \text{ m}, 120.0°)$

2. The cartesian coordinates x and y are related to the polar coordinates, r and θ

$$x = r \cos \theta \qquad \text{and} \qquad y = r \sin \theta$$

Solving the problem

3. For each vector, $\vec{A}$ and $\vec{B}$, the x and y components are

$A_x = (2.50 \text{ m}) \cos 30° = 2.17 \text{ m}$
$A_y = (2.50 \text{ m}) \sin 30° = 1.25 \text{ m}$

$B_x = (3.80 \text{ m}) \cos 120° = -1.90 \text{ m}$
$B_y = (3.80 \text{ m}) \sin 120° = 3.29 \text{ m}$

$\vec{A} = (2.17 \text{ m}, 1.25 \text{ m})$
$\vec{B} = (-1.90 \text{ m}, 3.29 \text{ m})$

4. The x and y coordinates of $\vec{B} - \vec{A}$ are

$$(\vec{B} - \vec{A})_x = -4.07 \text{ m} \qquad (\vec{B} - \vec{A})_y = 2.04 \text{ m}$$

5. The magnitude of $\vec{B} - \vec{A}$ is

$$|\vec{B} - \vec{A}| = \sqrt{4.07^2 + 2.04^2} = 4.55 \text{ m}$$

Before we begin...

1.

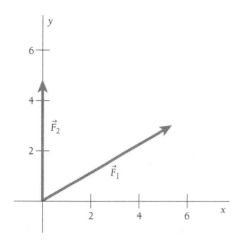

Solving the problem

2.

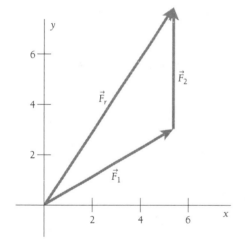

The resultant force is found by translating the tail of $\vec{F}_2$ to the head of $\vec{F}_1$. The resultant force $\vec{F}_r$ is 9.54 units directed 57° above the x axis.

Before we begin...

1. The rectangular coordinates of a vector are related to the polar coordinates by

$$A_x = A \cos \theta \quad \text{and} \quad A_y = A \sin \theta$$

2.

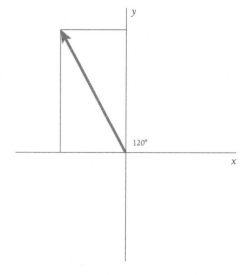

Solving the problem

3. The rectangular coordinates A_x and A_y are

$$A_x = (50.0 \text{ m}) \cos 120° = -25.0 \text{ m}$$

and

$$A_y = (50.0 \text{ m}) \sin 120° = 43.3 \text{ m}$$

Before we begin...

1.

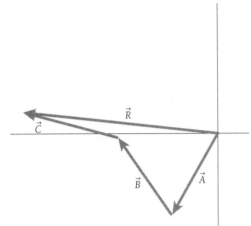

2. The vectors $\vec{A}$, $\vec{B}$, and $\vec{C}$ can be expressed in polar coordinates as $\vec{A} = (75\ \text{paces}, 240°)$, $\vec{B} = (125\ \text{paces}, 135°)$, and $\vec{C} = (100\ \text{paces}, 160°)$

Solving the problem

3. The x and y components of the three vectors are

$A_x = (75\ \text{paces}) \cos 240° = -37.5\ \text{paces}$
$A_y = (75\ \text{paces}) \sin 240° = -65.0\ \text{paces}$

$B_x = (125\ \text{paces}) \cos 135° = -88.4\ \text{paces}$
$B_y = (125\ \text{paces}) \sin 135° = 88.4\ \text{paces}$

$C_x = (100\ \text{paces}) \cos 160° = -94.0\ \text{paces}$
$C_y = (100\ \text{paces}) \cos 160° = 34.2\ \text{paces}$

4. Adding the x components together and the y components together yields

$R_x = -220\ \text{paces}$ and $R_y = 57.6\ \text{paces}$

5. The magnitude of the resultant vector $\vec{R}$ is

$$R = \sqrt{R_x^2 + R_y^2} = 227\ \text{paces}$$

6. The angle that the resultant makes with respect to the x axis is

$$\theta = \tan^{-1}\left(\frac{R_y}{R_x}\right) = -15°$$

Inspection of the signs of the x and y components of R shows that the angle is in the second quadrant, 15° short of the $-x$ axis. Therefore,

$$\theta = 165°$$

Before we begin...

1.

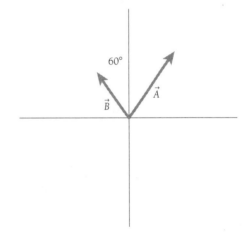

2. The magnitudes of the two vectors are

$|\vec{A}| = 7$ $\qquad$ $|\vec{B}| = 4$

3. The resultant of $\vec{A} \bullet \vec{B}$ is a scalar quantity.

Solving the problem

4. $\vec{A} \bullet \vec{B} = 28 \cos 60° = 14$

Solutions

PROBLEM 6 The Dot Product

Before we begin...

1. This problem expresses the vectors in unit vector notation or rectilinear coordinate form.

2. The x, y, and z components of the vectors are

$$A_x = 0 \qquad A_y = 2.0 \qquad A_z = 0$$

$$B_x = -5.0 \qquad B_y = 3.0 \qquad B_z = 0$$

Solving the problem

3. The scalar product is

$$\vec{A} \bullet \vec{B} = 0 + 6.0 + 0 = 6.0$$

The magnitudes of A and B are

$$A = 2.0 \quad \text{and} \quad B = \sqrt{3.0^2 + 5.0^2} = 5.8$$

4. The angle is

$$\theta = \cos^{-1}\left(\frac{6.0}{2.0 \times 5.8}\right) = 59°$$

PROBLEM 7 The Cross Product

Before we begin...

1. The components of the two vectors are

$$M_x = 6 \qquad M_y = 2 \qquad M_z = -1$$

$$N_x = 2 \qquad N_y = -1 \qquad N_z = -3$$

2. The resultant of the vector product is a vector quantity.

Solving the problem

3. Computing $\vec{M} \times \vec{N}$, we find

$$\vec{M} \times \vec{N} = (-6 - 1)\hat{i} - [-18 - (-2)]\hat{j}$$
$$+ (-6 - 4)\hat{k}$$
$$= -7\hat{i} + 16\hat{j} - 10\hat{k}$$

PROBLEM 8 Vector (Cross) Product

Before we begin...

1. The directions of the two vectors identify the plane in which they lie.

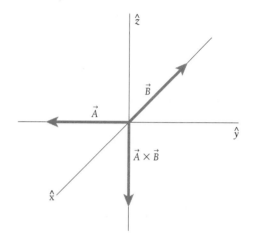

2. The vector product is perpendicular to the plane of the two vectors.

Solving the problem

3. The value of θ as measured counterclockwise (as per the right-hand rule) is 270°.

4. $\vec{A} \times \vec{B}$ is in the $-z$ direction.

The direction of $\vec{B} \times \vec{A}$ is the opposite of $\vec{A} \times \vec{B}$, or in the $+z$ direction.

Saunders Core Concepts in Physics Workbook

PROBLEM 9 Displacement, Velocity, and Speed

Before we begin...

1. The acceleration vector is the slope of the velocity as a function of time.

Solving the problem

2. The acceleration is constant from $t = 0$ to $t = 2$ s.

The acceleration in this region is

$$\text{slope} = \frac{(-3 - 0) \text{ m/s}}{(2 - 0) \text{ s}} = -1.5 \frac{\text{m}}{\text{s}^2}$$

3. The acceleration is constant from $t = 2$ to $t = 6$ s.

The value of the acceleration in this region is

$$\text{slope} = \frac{[3 - (-3)] \text{ m/s}}{(6 - 2) \text{ s}} = +1.5 \frac{\text{m}}{\text{s}^2}$$

4. The acceleration is constant from $t = 6$ to $t = 8$ s.

The value of the acceleration in this region is

$$\text{slope} = \frac{(3 - 3) \text{ m/s}}{(8 - 6) \text{ s}} = 0 \frac{\text{m}}{\text{s}^2}$$

5.

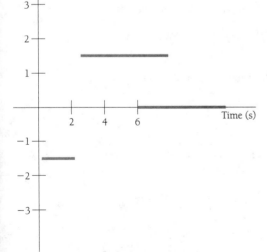

6. The average acceleration from $t = 2$ s to $t = 8$ s is

$$\langle \vec{a} \rangle = \frac{[3 - (-3)] \text{ m/s}}{(8 - 2) \text{ s}} = +1.0 \frac{\text{m}}{\text{s}^2}$$

7. At $t = 4.0$ s, the acceleration is $+1.5$ m/s^2

PROBLEM 10 Instantaneous Velocity and Acceleration

Before we begin...

1. Instantaneous velocity and acceleration are related to displacement as a function of time by the equations

$$\vec{v} = \frac{d\vec{y}}{dt} \qquad \vec{a} = \frac{d\vec{v}}{dt} = \frac{d^2\vec{y}}{dt^2}$$

Solving the problem

2. The speed is

$$v = [7.00 - (9.80)t] \text{ m/s}$$

therefore $v_0 = 7.00$ m/s.

3. The velocity at $t = 1.26$ s is

$$\vec{v} = 7.00 - (9.80)(1.26) = -5.35 \text{ m/s}$$

4. The acceleration is

$$\vec{a} = \frac{d\vec{v}}{dt} = -9.80 \text{ m/s}^2$$

solutions

PROBLEM 11 Motion in One Dimension

Before we begin...

1. Both players will travel the same distance.

2. The given information concerning the motion of each player is

First Player	Second Player
$v_{01} = 0$	$v_{02} = +12.0$ m/s
$a_1 = +4.0$ m/s^2	$a_2 = 0$

3. The first player catches the second player at

$$t_2 = t_1 + 3.0 \text{ s}$$

Solving the problem

4. An expression for the distance traveled by the first player as related to original velocity v_0, acceleration a, and time t_1 is

$$x = v_{01}t_1 + \tfrac{1}{2}a_1t_1{}^2$$

5. An expression for the distance traveled by the second player as related to his average velocity and time t_2 is

$$x = \langle v_2 \rangle t_2$$

6. The time required for the first player to overtake the second is

$$(12.0 \text{ m/s})(t_1 + 3 \text{ s}) = 0 + \tfrac{1}{2}(4.0 \text{ m/s}^2)t_1{}^2$$

$$t_1 = +8.2 \text{ s} \quad \text{or} \quad -2.2 \text{ s}$$

We select +8.2 s because the players should meet at some time in the future.

7. The distance that a player has traveled is

$$x_1 = 0 + \tfrac{1}{2}(4.0 \text{ m/s}^2)(8.2 \text{ s})^2 = 134 \text{ m}$$

Both players should travel the same distance.

PROBLEM 12 Projectile Motion

Before we begin...

1.

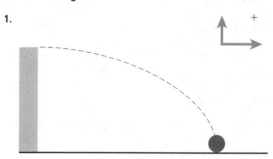

2. The given information is

Horizontal	Vertical
$\langle v_x \rangle = v_{0x} = ?$	$v_{0y} = 0$
$a_x = 0$	$a_y = -9.80$ m/s^2
$\Delta x = +80$ m	$\Delta y = -35$ m

3. Without air resistance, the horizontal velocity does not change.

Solving the problem

4. An equation of motion that relates time t to original velocity v_{0y}, acceleration a_y, and displacement Δy, is

$$\Delta y = v_{0y}t + \tfrac{1}{2}a_yt^2$$

5. Substituting and solving for the time of flight yields

$$t = +2.67 \text{ s}$$

6. The final vertical component of velocity is

$$v_{fy} = v_{0y} + a_yt$$

$$= -26.2 \text{ m/s}$$

7. The horizontal component of the velocity is found by

$$\langle v_x \rangle = \frac{\Delta x}{\Delta t} = +29.9 \text{ m/s}$$

Saunders Core Concepts in Physics Workbook

Before we begin...

1.

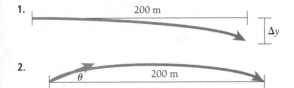

2.

3. The given information is

Horizontal	Vertical
$v_{0x} = 500$ m/s	$v_{0y} = 0$
$a_x = 0$	$a_y = -g = -9.80$ m/s^2
$\Delta x = 200$ m	

Solving the problem

4. Since in the horizontal direction $\langle v \rangle = \Delta x / \Delta t$, $\Delta t = (200$ m$)/(500$ m/s$) = 0.4$ s.

5. $\Delta y = v_{0y} + \frac{1}{2}a_y t^2 = -0.784$ m

6. To have the bullet hit the center of the target, $\sin (2\theta) = Rg/(v_0^2) = 0.00784$, so $\theta = \frac{1}{2} \sin^{-1} (0.00784) = 0.225°$ above the horizontal.

Before we begin...

1. This problem involves both projectile motion and uniform circular motion.

2. To compute centripetal acceleration, we must know the speed of the object and the radius of its path.

3. Projectile motion analysis will give us the speed of the ball as the string breaks.

4. The given information is

$$\Delta y = -1.2 \text{ m} \qquad \Delta x = +2.0 \text{ m}$$
$$a_y = -9.80 \text{ m/s}^2 \qquad v_{0y} = 0$$

Solving the problem

5. An equation of motion that relates Δy, v_{0y}, a_y and t is

$$\Delta y = v_{0y} + \frac{1}{2}a_y t^2$$

6. Substituting into the equation and solving for the time required for the ball to cover the vertical displacement yields

$$-1.2 \text{ m} = 0 + \frac{1}{2}(-9.8 \text{ m/s}^2)t^2 \qquad t = 0.49 \text{ s}$$

7. The x component of velocity required to cover the x displacement Δx is

$$\langle v_x \rangle = \Delta x / t = +4.04 \text{ m/s}$$

8. Using the x component of velocity as the speed of the ball when the string was still attached and computing the centripetal acceleration of the ball gives us

$$a_r = v^2/r = 54.4 \text{ m/s}^2$$

Solutions

Before we begin...

1. The centripetal acceleration and the tangential acceleration at any point in time are at right angles to each other.

2.

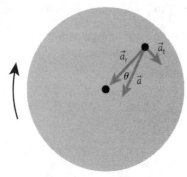

Rotating table

3. The given information for the tangential motion of the point in the problem is

$$v_0 = 0$$

$$v_f = 0.700 \text{ m/s}$$

$$\Delta t = 1.75 \text{ s}$$

Solving the problem

4. The equation of motion for uniformly accelerated tangential motion that relates original velocity, final velocity, time, and acceleration is

$$a_t = \Delta v / \Delta t$$

5. The tangential acceleration a_t is

$$a_t = (0.70 \text{ m/s} - 0)/1.75 \text{ s} = 0.4 \text{ m/s}^2$$

6. We use $t = 1.25$ s, not $t = 1.75$ s, because the velocity is changing during the time interval.

7. The tangential velocity is

$$v_f = v_0 + at$$
$$= 0 + (0.4 \text{ m/s}^2)(1.25 \text{ s}) = 0.50 \text{ m/s}$$

8. The centripetal acceleration is

$$a_r = v^2/r = (0.5 \text{ m/s})^2/0.2 \text{ m} = 1.25 \text{ m/s}^2$$

9. Using the Pythagorean theorem, the vector sum is

$$a = \sqrt{a_t^2 + a_r^2} = 1.31 \text{ m/s}^2$$

$$\theta = \tan^{-1}\left(\frac{a_t}{a_r}\right) = 17.8°$$

where θ is measured with respect to the centripetal acceleration.

Before we begin...

1.

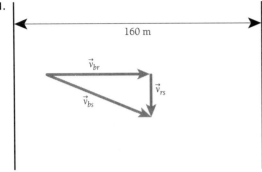

160 m

2. The vector sum is important because the current is pushing the boat downstream at the same time the boat is crossing the river.

Solving the problem

3. The magnitude of the velocity of the boat relative to the shore is

$$v_{bs} = \sqrt{v_{br}^2 + v_{rs}^2} = 2.5 \text{ m/s}$$

4. The time required to cross the river is

$$t = \Delta x / \langle v_x \rangle = (160 \text{ m})/(2.00 \text{ m/s}) = 80.0 \text{ s}$$

5. The distance downstream that the boat will travel is

$$\Delta y = \langle v_y \rangle t = (1.5 \text{ m/s})(80 \text{ s}) = 120 \text{ m}$$

Module 4 **Forces**

PROBLEM 17 Motion, Newton's First Law, and Force

Before we begin...

1. The net force being exerted on the boat must be zero if it moves at constant velocity.

2.

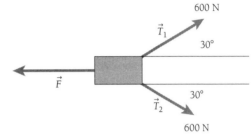

600 N

30°

30°

600 N

Solving the problem

3. $T_{1x} + T_{2x} + F_x = 0$, where $T_{1x} = T_1 \cos 30°$ and $T_{2x} = T_2 \sin 30°$, becomes 520 N + 520 N + F_x = 0; and $T_{1y} + T_{2y} + F_y = 0$, where $T_{1y} = T_1 \sin 30°$ and $T_{2y} = -T_2 \sin 30°$, becomes 300 N − 300 N + F_y = 0.

4. The x and y components of the resistive force $\vec{F}$ are

$$F_x = -1040 \text{ N} \quad \text{and} \quad F_y = 0$$

Therefore $\vec{F} = -1040\hat{i}$ N or 1040 N in the $-x$ direction.

Solutions

PROBLEM 18 Inertia, Mass, and Weight

Before we begin...

1. The acceleration is equal to the force divided by the mass.

Solving the problem

2. The magnitude of the force exerted by the Earth on the student is

$$\vec{F} = m\vec{g} = (60 \text{ kg})(9.80 \text{ m/s}^2) = 588 \text{ N}$$

toward the center of Earth.

3. Setting $M_E \, \vec{a}_E = 588$ N upward yields

$$\vec{a}_E = (588 \text{ N})/(5.98 \times 10^{24} \text{ kg})$$
$$= 9.83 \times 10^{-23} \text{ m/s}^2 \text{ upward}$$

PROBLEM 19 Newton's Second Law

Before we begin...

1. The given information for this problem is

$$\vec{v}_i = 3.0 \text{ m/s} \qquad m_1 = 85\text{-kg} \qquad \Delta t = 0.5 \text{ s}$$
$$\vec{v}_f = 4.0 \text{ m/s} \qquad m_2 = 58\text{-kg}$$

2. The acceleration of the sprinter and the force required to cause this acceleration need to be computed.

Solving the problem

3. The acceleration that the sprinter must experience is

$$\vec{a} = (\vec{v}_f - \vec{v}_i)/t = 2.0 \text{ m/s}^2$$

4. Newton's second law allows you to calculate the net force from the mass and the acceleration.

5. The force on the 85-kg sprinter is

$$\vec{F}_{net} = M\vec{a} = (85 \text{ kg})(2.0 \text{ m/s}^2) = 170 \text{ N}$$

6. This same force would cause the 58-kg sprinter to accelerate

$$\vec{a} = (170 \text{ N})(58 \text{ kg}) = 2.93 \text{ m/s}^2$$

PROBLEM 20 Newton's Third Law

Before we begin...

1.

2. All three blocks must have the same acceleration. The applied force is pushing on the first block.

Solving the problem

3. The net force on each block is related to the block's mass and acceleration as follows:

Block 1 $\qquad \vec{F} - \vec{P}_1 = M_1 \vec{a}$

Block 2 $\qquad \vec{P}_1 \quad \vec{P}_2 = M_2 \vec{a}$

Block 3 $\qquad \vec{P}_2 = M_3 \vec{a}$

4. Solving the three simultaneous equations for the acceleration yields

$$\vec{a} = 2.00 \text{ m/s}^2$$

5. The contact force acting upon the object is

Block 3 $\quad P_2 = (4.00 \text{ kg})(2.00 \text{ m/s}^2) = 8.00 \text{ N}$

Block 1 $\quad (18.0 \text{ N}) - P_1 = (2.00 \text{ kg})(2.00 \text{ m/s}^2)$
$$= 4.00 \text{ N}$$

$$P_1 = 14.0 \text{ N}$$

As a check, the net force on Block 2 must be 6.00 N. Observe that $P_1 - P_2 = 6.00$ N.

Saunders Core Concepts in Physics Workbook

PROBLEM 21 Free-Body Diagrams

Before we begin...

1.

2. The system is in equilibrium; the net force must be zero.

3. The weight of the ball is

$$W = mg = 98.0 \text{ N}$$

Solving the problem

4. The magnitude of the tension T_3 is

$$T_3 = W = 98.0 \text{ N}$$

5. The first free-body diagram shows that the tension $\vec{T}_3$ is pulling vertically downward ($-y$ direction). The directions of $\vec{T}_1$ and $\vec{T}_2$ are

$$T_2 \text{ is at } 0° \text{ and } T_1 \text{ is at } 120°$$

6. Resolving the tensions into their x and y components yields

$$T_{1x} = T_1 \cos 120° = -0.5 \, T_1$$
$$T_{1y} = T_1 \sin 120° = +0.866 \, T_1$$

$$T_{2x} = T_2 \cos 0° = T_2$$
$$T_{2y} = T_2 \sin 0° = 0$$

$$T_{3x} = T_3 \cos 270° = 0$$
$$T_{3y} = T_3 \sin 270° = -98.0 \text{ N}$$

Solving the two equations of equilibrium simultaneously yields

$$T_1 = 113 \text{ N} \quad \text{and} \quad T_2 = 56.5 \text{ N}$$

PROBLEM 22 Free-Body Diagrams

Before we begin...

1. The 4.0-kg object weighs more and so will exert more downward force than the 2.0-kg object. By common sense, the 4.0-kg object will accelerate downward.

2. All of the accelerations must be the same because the strings are assumed not to stretch.

3.

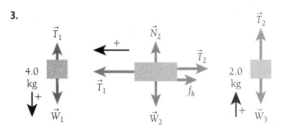

Solving the problem

4. For the three objects, we find

$$4.0 \text{ kg} \qquad W_1 - T_1 = M_1 a$$

$$1.0 \text{ kg} \qquad T_1 - T_2 = M_2 a$$

$$2.0 \text{ kg} \qquad T_2 - W_3 = M_3 a$$

5. Solving the first and third equations for T_1 and T_2, and substituting these results into the second equation, we obtain

$$a = \frac{W_1 - W_3}{M_1 + M_2 + M_3} = \frac{39.2 \text{ N} - 19.6 \text{ N}}{7 \text{ kg}} = 2.8 \text{ m/s}^2$$

6. Substituting this result back into the first and third equations yields

$$T_1 = W_1 - M_1 a$$
$$= 39.2 \text{ N} - (4 \text{ kg})(2.8 \text{ m/s}^2) = 28.0 \text{ N}$$

$$T_2 = W_2 + M_2 a$$
$$= 19.6 \text{ N} + (2 \text{ kg})(2.8 \text{ m/s}^2) = 25.2 \text{ N}$$

As a consistency check, note that this gives a net force on the 1-kg mass of $T_1 - T_2 = 2.8 \text{ N}$. This is just what is needed to give it the predicted acceleration of 2.8 m/s^2, so our result is consistent.

PROBLEM 23 Centripetal Force

Before we begin...

1.

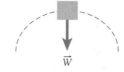

2. A normal force does not exist between the driver and the seat because the weight of the driver is providing exactly the centripetal force.

Solving the problem

3. The net force providing the centripetal force is

$$W = mg = F_c = \frac{mv^2}{r}$$

4. The speed of the car is

$$v = \sqrt{rg} = 13.3 \text{ m/s}$$

5. The mass appears in both the net force and the centripetal force. Both sides of the equation can be divided by the mass without altering the relationship.

PROBLEM 24 Fictitious Forces: Motion in Accelerated Reference Frames

Before we begin...

1.

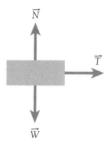

Solving the problem

2. The acceleration of the mass with respect to the inertial frame of reference is

$$\vec{T} = m\vec{a}, \quad \text{so}$$
$$\vec{a} = \frac{\vec{T}}{m} = \frac{18.0 \text{ N}}{5.00 \text{ kg}} = 3.60 \text{ m/s}^2$$

3. When $\vec{a} = 0$, $\vec{T}$ also must be zero.

4. In the noninertial frame of reference, the net force appears to be zero. For this to be accomplished, we must introduce a fictitious force acting to the left and equal in magnitude to the tension $\vec{T}$.

PROBLEM 25 Work

Before we begin...

1. The work done by a variable force is calculated by

$$W = \int_{s_1}^{s_2} \vec{F} \cdot d\vec{s}$$

2. The displacement $d\vec{s}$ is in the positive x direction; $d\vec{s} = dx\hat{i}$.

Solving the problem

3. The integral expression for the work done gives us the following:

$$W = \int_{x=0}^{x=5.0\ m} (4.0x\hat{i} + 3.0y\hat{j})\ N \cdot dx\hat{i}$$

Therefore,

$$W = \int_{x=0}^{x=5.0\ m} (4.0x\hat{i})\ N \cdot dx\hat{i} + 0$$

or

$$W = (4\ N/m)\frac{x^2}{2}\Big|_{0}^{5\ m} = 50.0\ J$$

PROBLEM 26 Important Examples of Work: Gravity and Springs

Before we begin...

1. The cheerleader has to exert a force equal to the weight of his partner.

$$\vec{F} = -\vec{W} = mg = (50.0\ kg)(9.80\ m/s^2)$$
$$= 490\ N\hat{j}$$

2. The displacement is $\Delta\vec{y} = 0.60\ m\hat{j}$.

Solving the problem

3. The work done for each replication is

$$W = \vec{F} \cdot \vec{s} = Fs\cos\theta$$

Because the force and the displacement are in the same direction, $\cos\theta = 1.00$. For each time the partner was lifted,

$$W = (490\ N)(0.60\ m) = 294\ J$$

4. The total work done is $20 \times 294\ J = 5.88 \times 10^3\ J$.

Solutions

PROBLEM 27 Important Examples of Work

Before we begin...

1. The limits of the integration are from $x = 0$ to $x = 0.10$ m

Solving the problem

2. Evaluating the integral between the limits and substituting the value of k into the expression yields

$$W = \int(-kx)dx$$

$$= \frac{-kx^2}{2} = \frac{(10 \text{ N/m})(0.1 \text{ m})^2}{2}$$

$$= -0.5 \text{ J}$$

PROBLEM 28 Work Done to Accelerate a Mass

Before we begin...

1. The units of the coefficients of the three terms are

$$15{,}000 \text{ N} \qquad 10{,}000 \text{ N/m} \qquad 25{,}000 \text{ N/m}^2$$

2. The expression for calculating the work done by a variable force is

$$W = \int_{s_1}^{s_2} \vec{F} \cdot d\vec{s}$$

Solving the problem

3. Substituting the given force into the equation for determining work done by a variable force yields

$$W = \int_{x_1}^{x_2} \vec{F} \cdot d\vec{x} = \int_{0}^{0.6 \text{ m}} (15{,}000 \text{ N} + 10{,}000x \text{ N/m}$$
$$- 25{,}000x^2 \text{ N/m}^2)dx$$

4. The integral between the limits of $x = 0$ and $x = 0.6$ m is

$$W = 9.00 \times 10^3 \text{ J}$$

5. Repeating between the limits of $x = 0$ and $x = 1.00$ m yields

$$W = 1.17 \times 10^4 \text{ J}$$

The last value is 30 percent greater.

Before we begin...

1. Conservative forces and the law of conservation of energy are used to solve this problem.

Solving the problem

2. The potential energy function can be evaluated by integrating the force over the interval of the displacement:

$$U = -\int_0^x (-Ax + Bx^2)\,dx$$

$$= \frac{Ax^2}{2} - \frac{Bx^3}{3}$$

3. Going from $x = 2.0$ m to $x = 3.0$ m yields

$$\Delta U = \frac{A(3^2 - 2^2)}{2} - \frac{B(3^3 - 2^3)}{3} = \frac{5}{2}A - \frac{19}{3}B$$

Because $\Delta K = -\Delta U$,

$$\Delta K = -\frac{5}{2}A + \frac{19}{3}B$$

Before we begin...

1. The given information is

$$m = 15.0 \text{ g} = 0.015 \text{ kg}$$

$$\vec{v}_0 = 0$$

$$\vec{v}_f = 780 \text{ m/s}$$

2. The work done by the net force acting upon a system is equal to the change in kinetic energy of the system.

3. The force accelerates the bullet over $\Delta \vec{x} = 72$ cm $= 0.72$ m.

Solving the problem

4. The change in kinetic energy of the bullet is

$$\Delta K = \frac{1}{2}mv_f^2 - \frac{1}{2}mv_0^2$$
$$= (0.5)(0.015 \text{ kg})(780 \text{ m/s})^2 - 0$$
$$= 4560 \text{ J}$$

5. The definition of work is used to calculate the average force:

$$W = \langle F \rangle \Delta x = 4560 \text{ J}$$

$$\langle F \rangle = 6330 \text{ N}$$

PROBLEM 31 Power

Before we begin...

1.

2. The given information is

$$m = 650 \text{ kg} \qquad v_0 = 0$$

$$v_f = 1.75 \text{ m/s} \qquad \Delta t = 3.00 \text{ s}$$

3. Kinetic energy and gravitational potential energy will be changed by the work done by the elevator motor.

Solving the problem

4. The change in kinetic energy is

$$\Delta K = \tfrac{1}{2}mv_f^2 - \tfrac{1}{2}mv_0^2 = 995 \text{ J}$$

5. The height to which the elevator rises in 3.00 s is

$$\Delta y = \langle v \rangle \Delta t = \frac{(0 + 1.75 \text{ m/s})}{2}(3.00 \text{ s}) = 2.63 \text{ m}$$

The change in potential energy is therefore

$$\Delta U = mg\Delta y = (650 \text{ kg})(9.80 \text{ m/s}^2)(2.63 \text{ m})$$
$$= 16{,}750 \text{ J}$$

The total work done by the motor in the first three seconds is 1.77×10^4 J.

6. The average power is calculated as the work done by the motor divided by the time required:

$$P = 5910 \text{ W}$$

7. Once the elevator is moving at a constant speed, the net force must be zero, so the force must equal the weight of the elevator.

Using $P = \vec{F} \cdot \langle \vec{v} \rangle$ after the elevator has reached its constant velocity,

$$P = (650 \text{ kg})(9.80 \text{ m/s}^2)(1.75 \text{ m/s})$$
$$= 1.11 \times 10^4 \text{ W}$$

PROBLEM 32 Conservation of Energy

Before we begin...

1. The law of conservation of energy states that the total mechanical energy of a system (kinetic energy + potential energy) must remain constant in any isolated system of objects that interact only through conservative forces.

2. The given information for the first question is

$$K_i = 30 \text{ J} \qquad \Sigma U_i = 10 \text{ J} \qquad K_f = 18 \text{ J}$$

3. For the second question,

$$\sum U_f = 5 \text{ J}$$

Solving the problem

4. Applying the law of conservation of energy for a conservative system:

$$K_i + \sum U_i = K_f + \sum U_f$$
$$30 \text{ J} + 10 \text{ J} = 18 \text{ J} + \sum U_f$$
$$\sum U_f = 22 \text{ J}$$

5. The total energy E at time t_i is

$$K_i + \sum U_i = 40 \text{ J}$$

6. In the second question, if $\sum U_f = 5$ J, then the total kinetic plus potential energy is no longer 40 J. Work totaling 17 J must have been done by a non-conservative force.

Saunders Core Concepts in Physics Workbook

PROBLEM 33 The General Form of Newton's
Second Law

Before we begin...

1. The general form of Newton's second law is expressed by the equation

$$\vec{F} = \frac{d\vec{p}}{dt} = m\frac{d\vec{v}}{dt} + \frac{dm}{dt}\vec{v}_{\text{rel}}$$

where $\vec{p}$ is the momentum of the system, $\vec{v}$ is the velocity of the system, and $\vec{v}_{\text{rel}}$ is the velocity with which mass is ejected.

2. Thrust is the force exerted on the object by the ejected mass. In this case, the ejected mass is the gas. Thrust is computed by

$$T = v_g\frac{dM_g}{dt}$$

3. The given information is

$$\frac{dM_g}{dt} = 80 \text{ kg/s} \qquad v_g = 2.5 \times 10^3 \text{ m/s}$$

Solving the problem

4. We can evaluate the thrust as follows:

$$T = v_g\frac{dM_g}{dt} = (2.5 \times 10^3 \text{ m/s})(80 \text{ kg/s})$$

$$= 2.0 \times 10^5 \text{ N}$$

Before we begin...

1. The given information is

$$m = 2.0 \text{ kg}$$

$$v_{i(\text{particle initially at rest})} = 0$$

$$v_{i(\text{particle initially moving})} = -2.0 \text{ m/s}$$

2. Impulse is related to force and the time interval over which the force is applied by the equation

$$\Delta\vec{p} = \vec{F}\Delta t$$

3. Momentum is defined as $\vec{p} = m\vec{v}$.

Solving the problem

4. The three regions are two triangles and one rectangle with respective areas of

$$4 \text{ N·s, } 4 \text{ N·s and } 4 \text{ N·s}$$

The total impulse therefore is 12 N·s.

5. Final momentum can be expressed in terms of the initial momentum and the impulse as

$$\vec{p}_f = \vec{p}_i + \Delta\vec{p}$$

6. For $\vec{v}_i = 0$, $\vec{p}_f = 0 + 12 \text{ N·s} = 12 \text{ N·s}$. Since $\vec{p} = m\vec{v}$, $\vec{v}_f = 6.0 \text{ m/s}$. If $\vec{v}_1 = -2.0 \text{ m/s}$, $\vec{p}_f = -4 \text{ N·s} + 12 \text{ N·s} = 8 \text{ N·s}$. Then, $\vec{v}_f = 4.0 \text{ m/s}$.

7. A constant force required to give the same value of impulse is

$$12.0 \text{ N·s} = \langle\vec{F}\rangle(5.0 \text{ s})$$

so

$$\langle\vec{F}\rangle = 2.40 \text{ N}$$

Solutions

PROBLEM 35 Perfectly Inelastic Collisions

Before we begin...

1.

2. The given information is

$$m_1 = 90 \text{ kg} \qquad \vec{v}_1 = +10 \text{ m/s}$$

$$m_2 = 120 \text{ kg} \qquad \vec{v}_2 = -4.0 \text{ m/s}$$

3. The two masses will stick and move as one.

4. Momentum is always conserved in a collision in an isolated system.

Solving the problem

5. Evaluating the total momentum of the system before the collision, we find

$$\vec{p}_T = \vec{p}_1 + \vec{p}_2 = (90 \text{ kg})(+10 \text{ m/s})$$
$$+ (120 \text{ kg})(-4.0 \text{ m/s})$$
$$= +420 \text{ N·s.}$$

6. Because the momentum after the collision is the same as before the collision, the velocity of the players (who stick together and move as one mass) can be evaluated as

$$\vec{p} = (m_1 + m_2)\vec{v}_c$$

$$\vec{v}_c = (+420 \text{ N·s})/(210 \text{ kg}) = +2.0 \text{ m/s}$$

Because the + direction was selected as the direction of the halfback's velocity (north) and because the final velocity is also positive, the players are moving north after the collision.

7. Before the collision, the total kinetic energy was

$$K_1 + K_2 = \tfrac{1}{2}m_1v_1^2 + \tfrac{1}{2}m_2v_2^2 = 5460 \text{ J}$$

After the collision, it becomes

$$K_C = \tfrac{1}{2}(m_1 + m_2)v_c^2 = 420 \text{ J}$$

The work that the players did on each other during the collision converted much of their kinetic energy into other forms. Most of this energy ends up as heat, while smaller amounts end up as sound waves and as lingering vibrations in the players' helmets or skeletal systems.

PROBLEM 36 Perfectly Inelastic Collisions

Before we begin...

1.

2. The given information is

$$m_1 = 5 \text{ g} \qquad \vec{v}_1 = (250 \text{ m/s}, 20°)$$

$$m_2 = 3 \text{ g} \qquad \vec{v}_2 = (280 \text{ m/s}, 165°)$$

3. The two bullets undergo a perfectly inelastic collision.

4. The problem must be solved in two dimensions because the motion is in two dimensions.

Solving the problem

5. The x and y components of momentum for each bullet before the collision are

$$p_{1x} = m_1v_1 \cos \theta_1 = 1175 \text{ g·m/s}$$

$$p_{1y} = m_1v_1 \sin \theta_1 = 428 \text{ g·m/s}$$

$$p_{2x} = m_2v_2 \cos \theta_2 = -811 \text{ g·m/s}$$

$$p_{2y} = m_2v_2 \sin \theta_2 = 217 \text{ g·m/s}$$

6. Solving for v_x,

$$(m_1 + m_2)v_x = p_{1x} = p_{2x}$$

$$v_x = \frac{(p_{1x} + p_{2x})}{(m_1 + m_2)} = \frac{364 \text{ g·m/s}}{8 \text{ g}} = 45.5 \text{ m/s}$$

7. For the y component we find

$$v_y = \frac{(p_{1x} + p_{2x})}{(m_1 + m_2)} = \frac{645 \text{ g·m/s}}{8 \text{ g}} = 80.6 \text{ m/s}$$

8. Evaluating to find the velocity of the combined mass after the collision, we find

$$v_c = \sqrt{v_x^2 + v_y^2} = 92.5 \text{ m/s}$$

$$\theta = \tan^{-1}\left(\frac{v_y}{v_x}\right) = 60.6°$$

Saunders Core Concepts in Physics Workbook

PROBLEM 37 Perfectly Elastic Collisions

Before we begin...

1. The given information is

$$m_1 = 2.00 \text{ kg} \quad m_2 = 4.00 \text{ kg} \quad \Delta h = -5 \text{ m}$$

$$v_1 = 0 \qquad\qquad v_2 = 0$$

2. Potential energy is converted into kinetic energy.

3. Linear momentum and kinetic energy are both conserved in an elastic collision.

Solving the problem

4. Using the law of conservation of energy to solve for the speeds of the two objects immediately before the collision,

$$\Delta U + \Delta K = 0 \text{ for each object}$$

(No nonconservative forces are present.)

$$mg\Delta h + (\tfrac{1}{2}mv^2 - 0) = 0$$

The mass cancels in the equation, so solving for the speed of each object gives

$$v = \sqrt{2gh} = 9.90 \text{ m/s}$$

From this, we conclude $v_1 = +9.90$ m/s and $v_2 = -9.90$ m/s immediately before the collision.

5. The total linear momentum and kinetic energy are

$$p_T = (2.00 \text{ kg})(+9.90 \text{ m/s})$$
$$+ (4.00 \text{ kg})(-9.90 \text{ m/s}) = -19.8 \text{ N} \cdot \text{s}$$

$$K_T = \tfrac{1}{2} (2.00 \text{ kg})(9.90 \text{ m/s})^2$$
$$+ \tfrac{1}{2} (4.00 \text{ kg})(9.90 \text{ m/s})^2 = 294 \text{ J}$$

hence, after the collision,

$$2v_1 + 4v_2 = -19.8 \qquad \text{and}$$

$$\tfrac{1}{2}(2)v_1^2 + \tfrac{1}{2}(4)v_2^2 = 294$$

(The units have been dropped, but will yield velocities in m/s.)

Solving for v_1 yields two possible solutions:

$$v_1 = +9.90 \text{ m/s} \qquad \text{and} \qquad v_1 = -16.5 \text{ m/s}$$

The first solution is rejected; it describes the condition before the collision. The appropriate solution is $v_1 = -16.5$ m/s.

Substitution back into the momentum equation with the value of v_1 yields

$$v_2 = +3.3 \text{ m/s}$$

6. The height to which each block will rise is

$$(0 - \tfrac{1}{2}m_1v_1^2) + (m_1gh_1 - 0) = 0$$

When we substitute the speed of the object after the collision, the height is

$$h_1 = \frac{v_1^2}{2g} = \frac{(-16.9 \text{ m/s})^2}{2(9.8 \text{ m/s}^2)} = 13.9 \text{ m}$$

Likewise, for object m_2, we obtain

$$h_2 = 0.56 \text{ m}$$

PROBLEM 38 Center of Mass

Before we begin...

1. Equal mass is located above and below the x axis at the same distance from the axis.

2. Based on symmetry, the center of mass will lie on the x axis; $y_{CM} = 0$.

Solving the problem

3. Resolving the distance of the hydrogen atoms to the oxygen atom, we find the x component

$$x = L \cos \theta = (0.100 \text{ nm}) \cos 53°$$
$$= 0.0602 \text{ nm}$$

4. Using the equation for the location of the x_{CM} for a set of discrete particles, we find

$$x_{CM} = \frac{\sum m_i x_i}{\sum m_i}$$

$$= \frac{(15.99 \text{ u})(0)}{+ (1.008 \text{ u})(0.0602 \text{ nm})}$$
$$\frac{+ (1.008 \text{ u})(0.0602 \text{ nm})}{(15.999 + 1.008 + 1.008) \text{ u}}$$

$$= 0.00673 \text{ nm}$$

The coordinates of the center of mass are (0.00673, 0) nm from the oxygen nucleus.

PROBLEM 39 Motion of a System of Particles

Before we begin...

1. The given information is

$$m_1 = 2.0 \text{ kg} \qquad \vec{v}_1 = (2.0\hat{i} - 10t\hat{j}) \text{ m/s}$$
$$m_2 = 3.0 \text{ kg} \qquad \vec{v}_2 = 4.0\hat{i} \text{ m/s}$$

2. The linear momentum of the center of mass of a system is equal to the total linear momentum of the system.

3. Acceleration is the time rate of change of velocity.

Solving the problem

4. From the definition of momentum of the center of mass, we have

$$\vec{p}_{CM} = M\vec{v}_{CM} = m_1\vec{v}_1 + m_2\vec{v}_2$$
$$= (2.0 \text{ kg})[(2.0\hat{i} - 10t\hat{j}) \text{ m/s}]$$
$$+ (3.0 \text{ kg})[(2.0\hat{i} - 10t\hat{j}) \text{ m/s}]$$
$$- (16\hat{i} - 20t\hat{j}) \text{ N·s}$$

5. Solving for the velocity of the center of mass, we find

$$\vec{v}_{CM} = \frac{(16\hat{i} - 20t\hat{j}) \text{ N·s}}{(2.0 \text{ kg} + 3.0 \text{ kg})} = (3.2\hat{i} - 4.0t\hat{j}) \text{ m/s}$$

6. At $t = 0.5$ s, $v_{CM} = (3.2\hat{i} - 2.0\hat{j})$ m/s.

7. The acceleration of the center of mass is

$$\vec{a}_{CM} = \frac{d\vec{v}_{CM}}{dt} = -4.0\hat{j} \text{ m/s}^2$$

8. The total momentum of the system is

$$\vec{p}_{tot} = \vec{p}_{CM} = M\vec{v}_{CM}$$
$$= 5.0 \text{ kg} [(3.21\hat{i} - 2.0\hat{j}) \text{ m/s}]$$
$$= (16\hat{i} - 10\hat{j}) \text{ N·s}$$

Before we begin...

1. The given information is

$M = 8.0$ kg $m = 2.0$ kg

$L = 6$ m $h = 2$ m

2. Because there is no external force with an x component acting on the system, x_{CM} will not move. This tells us that the momentum of the center of mass $\Delta p_{CM(x)}$ of the system was and will remain at rest.

Solving the problem

3. The expression for dm in terms of the surface mass density, M/A, and the area of the strip ydx is

$dm = (M/A)ydx$

4. Evaluating the x_{CM}, we obtain

$$x_{CM} = \frac{1}{M} \int x \frac{M}{A} ydx$$

From similar triangles $y/x = h/L$, so that $y = (h/L)x$. Also recall that $A = \frac{1}{2} Lh$.

Substituting into the expression for the x_{CM}, we get

$$x_{CM} = \frac{1}{M} \int x \frac{M}{1/2hL}(h/L)xdx = \frac{2}{L^2} \int x^2 dx = \frac{2L}{3}$$

5. Computing the location of the center of mass of the total system by placing M at the x_{CM} of the triangle and m at the position $x = L$ yields

$$x_{CM} = \frac{(8.0 \text{ kg})(4 \text{ m}) + (2.0 \text{ kg})(6 \text{ m})}{8 \text{ kg} + 2 \text{ kg}} = 4.4 \text{ m}$$

6. Setting the value calculated equal to the situation where mass m is located at d and M is located at $(x_{CM \text{ triangle}} + d)$ gives

$$\frac{(8.0 \text{ kg})(4 \text{ m} + d) + (2.0 \text{ kg})d}{(8 \text{ kg} + 2 \text{ kg})} = 4.4 \text{ m}$$

Solving for d, we find

$d = 1.20$ m

Before we begin...

1. The given information is

$v_i = 0$ $v_f = 25$ m/s $\theta = 1.25$ rev $R = 1.00$ m

2. Angular speed is related to tangential or linear speed by the equation

$$\omega = \frac{v}{R}$$

Solving the problem

3. Converting the initial and final speeds to angular speeds using the linear transformation equations yields

$\omega_i = 0$ $\omega_f = 25.0$ rad/s

4. An equation from rotational kinematics that involves initial and final angular speeds, the angle through which the acceleration occurs, and the angular acceleration is

$$2\alpha\theta = \omega_f{}^2 - \omega_i{}^2$$

5. Converting the angle from revolutions to radians gives

$$\theta = (1.25 \text{ rev})(2\pi \text{ rad/rev}) = 7.85 \text{ rad}$$

6. Solving for angular acceleration we find

$$\alpha = \frac{\omega_f{}^2 - \omega_i{}^2}{2\theta} = \frac{(25.0 \text{ rad/s})^2}{2(7.85 \text{ rad})} = 39.8 \text{ rad/s}^2$$

7. To solve for the time, use

$$\alpha = \frac{\Delta\omega}{\Delta t}$$

$$\Delta t = \frac{\Delta\omega}{\alpha} = \frac{25.0 \text{ rad/s}}{39.8 \text{ rad/s}^2} = 0.628 \text{ s}$$

Solutions

PROBLEM 42 Rotational Energy

Before we begin...

1. The given information is

$$m = 215 \text{ g} = 0.215 \text{ kg} \qquad r = d/2 = 0.0319 \text{ m}$$

$$\Delta x = 3.00 \text{m} \qquad\qquad \theta = 25°$$

$$\Delta t = 1.50 \text{ s} \qquad\qquad v_i = 0$$

2. The length of the cylinder does not matter as long as it is rolling on its side.

3. The total energy of the system at the top of the incline will be equal to the total energy when the cylinder reaches the bottom.

Solving the problem

4. Computing ω_f, we find

$$\langle v \rangle = \Delta x/\Delta t \qquad \text{and} \qquad \langle v \rangle = {}^1\!/_2(v_i + v_f)$$

so

$$v_i + v_f = 2\Delta x/\Delta t$$

$$v_f = 4.00 \text{ m/s}$$

from which

$$\omega_f = v_f/r = 125 \text{ rad/s}$$

5. The initial gravitational potential energy is

$$U_i = mgh$$
$$= (0.215 \text{ kg})(9.80 \text{ m/s}^2)[(3 \text{ m}) \sin 25°]$$

6. Setting the total energy at the top of the incline equal to the total energy at the bottom, we obtain

$$(K_t + K_r + U)_{\text{top}} = (K_t + K_r + U)_{\text{bottom}}$$

$$0 + 0 + mgh = {}^1\!/_2 mv^2 + {}^1\!/_2 I\omega^2 + 0$$

We can then solve for I:

$$I = \frac{m(2gh + v^2)}{\omega^2} = 1.21 \times 10^{-4} \text{ kg·m}^2$$

PROBLEM 43 Moment of Inertia of Rigid Bodies

Before we begin...

1. The moment of inertia of a point mass rotating about a fixed axis is computed by $I = mr^2$, where r is the distance from the point of rotation to the mass m.

Solving the problem

2. The equation for the total moment of inertia of the two-object system is

$$I = Mx^2 + m(L - x)^2$$

3. Taking the first derivative of the function and setting it equal to zero to find the following values of x that satisfy the equation,

$$\frac{dI}{dx} = 2Mx - 2m(L - x) = 0$$

$$x = \frac{mL}{m + M}$$

4. By determining that

$$\frac{d^2I}{dx^2} = 2m + 2M$$

is positive, we see that the value of x is a minimum.

5. The center of mass of the system can be calculated according to the methods of Module 6, Linear Momentum, and shown to be the same as the value of x in the solution for question 3.

$$x_{\text{CM}} = \frac{\sum x_i m_i}{\sum m_i} = \frac{mL + M(0)}{(m + M)} = \frac{mL}{(m + M)}$$

Calculating I_{CM}, we find

$$I_{\text{CM}} = M\left(\frac{mL}{m + M}\right)^2 + m\left(L - \frac{mL}{m + M}\right)^2$$

Simplifying this expression yields

$$I_{\text{CM}} = \left(\frac{Mm}{m + M}\right)L^2 = \mu L^2$$

Before we begin...

1. The given information in unit vector notation is

$$\vec{r} = (\hat{i} + 3\hat{j}) \text{ m}$$

$$\vec{F} = (3\hat{i} + 2\hat{j}) \text{ N}$$

2.

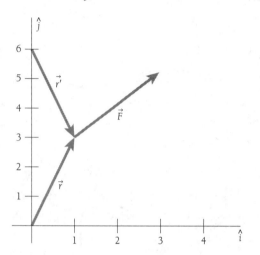

3. Torque is related to position and applied force by

$$\tau = \vec{r} \times \vec{F}$$

Solving the problem

4. We can use the determinant form to write the expression for $\tau = \vec{r} \times \vec{F}$ and solve for the torque:

$$\vec{r} \times \vec{F} = \begin{vmatrix} \hat{i} & \hat{j} & \hat{k} \\ 1 & 3 & 0 \\ 3 & 2 & 0 \end{vmatrix} = (2 - 9)\hat{k} = -7\hat{k} \text{ N·m}$$

5. The position vector in terms of the new point of rotation is

$$\vec{r}' = (1, 3, 0) - (0, 6, 0) = (1, -3, 0) \quad \text{or}$$

$$\vec{r}' = (\hat{i} - 3\hat{j}) \text{ m}$$

6. Solving for the torque about the new point of rotation yields

$$\vec{r}' \times \vec{F} = \begin{vmatrix} \hat{i} & \hat{j} & \hat{k} \\ 1 & -3 & 0 \\ 3 & 2 & 0 \end{vmatrix} = (2 + 9)\hat{k} = 11\hat{k} \text{ N·m}$$

7. It should be noted that $\vec{r}$ and $\vec{F}$ are both in the xy plane. The torque vectors are perpendicular to the r, F plane. This places them on either the z or $-z$ axis.

Solutions

Work and Energy Problem

Before we begin...

1. The given information is

$M = 100$ kg $R = 0.5$ m $F_n = 70$ N

$\Delta t = 6.0$ s $\omega_i = 50$ rev/min

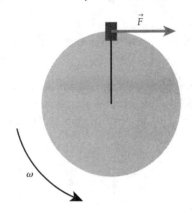

2. The frictional torque opposes the direction of rotation and thus decreases the rotational speed of the wheel.

3. The final rotational kinetic energy of the wheel will be zero.

Solving the problem

4. Using the equation $I = \frac{1}{2}MR^2$ for a solid disk, we calculate the moment of inertia of the wheel to be

$I = \frac{1}{2}(100 \text{ kg})(0.5 \text{ m})^2 = 12.5$ kg·m^2

5. The angular speed is converted to

$\omega_i = (50 \text{ rev/min})(2\,\pi \text{ rad/rev})(1 \text{ min/60 s})$
$= 5.23$ rad/s

This must be done so that the expression will be in fundamental units.

6. The initial rotational kinetic energy and the change in kinetic energy during the problem are

$K_i = \frac{1}{2}I\omega_i^2 = 171$ J

$\Delta K = (0 - 171 \text{ J}) = -171$ J

7. The formula for the work done by a constant net torque is

$\tau_{net}\Delta\theta = \Delta K$

8. To find $\Delta\theta$,

$(\omega_i + \omega_f)/2 = \Delta\theta\Delta/t$ so $\Delta\theta = 15.7$ rad

9. Using the work-energy theorem to solve for the torque,

$(\tau_{net})(15.7 \text{ rad}) = 171$ J

$\tau_{net} = 10.9$ N·m

(Note that the negative sign indicates the direction of the torque and is therefore ignored in computing the magnitude.)

PROBLEM 46 Rolling Motion

Before we begin...

1.

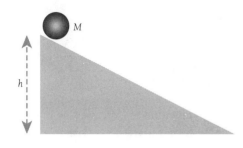

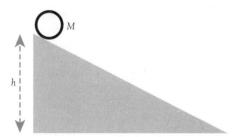

2. The moments of inertia for the objects are

$$I_{disk} = \frac{1}{2}MR^2 \qquad I_{hoop} = MR^2$$

3. Each object has potential energy at the top of the incline.

Solving the problem

4. A rolling object has both rotational kinetic energy and translational kinetic energy. The expressions for the type of kinetic energy are

$$K_{rot} = \frac{1}{2}I\omega^2 \qquad K_{tran} = \frac{1}{2}Mv^2$$

5. The law of conservation of energy as it applies to this problem is

$$U_i + K_{i\text{-rot}} + K_{i\text{-tran}} = U_f + K_{f\text{-rot}} + K_{f\text{-tran}}$$

Substituting the known quantities into this general equation,

$$Mgh + 0 + 0 = 0 + \frac{1}{2}I\omega_f^2 + \frac{1}{2}Mv_f^2$$

For a disk,

$$Mgh + 0 + 0 = 0 + \frac{1}{2}(\frac{1}{2}MR^2)\omega_f^2 + \frac{1}{2}Mv_f^2$$

For a hoop,

$$Mgh + 0 + 0 = 0 + \frac{1}{2}(MR^2)\omega_f^2 + \frac{1}{2}Mv_f^2$$

Using $v = R\omega$, we can solve after substitution

For a disk,

$$v = \sqrt{\frac{4}{3}gh}$$

For a hoop,

$$v = \sqrt{gh}$$

6. The disk has the greater speed at the bottom.

Because the objects cover the same distance, the one with the greater average speed will reach the bottom first. Because they have the same initial speed, the one with the greater final speed will have the greater average speed as well.

7. The disk stores less of its kinetic energy in rotation, leaving more for translational kinetic energy.

Before we begin...

1. The angular momentum of a particle rotating in a constant circular path is computed by $L = \vec{R} \times \vec{p}$ where $\vec{p} = m\vec{v}$. $L = Rmv \sin \theta$, where θ is the angle between $\vec{R}$ and $\vec{p}$.

2.

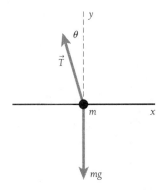

3. A particle moving in a constant circular path must be subjected to a net centripetal force.

Solving the problem

4. The tension $\vec{T}$ is resolved into its x and y components:

$$T_x = T \sin \theta \quad \text{and} \quad T_y = T \cos \theta$$

For the x forces,

$$T \sin \theta = ma = mv^2/R$$

In the y direction,

$$T \cos \theta - mg = 0$$

5. Solving these equations for T and equating the expressions,

$$\frac{mg}{\cos \theta} = \frac{mv^2}{R \sin \theta}$$

so,

$$v = \sqrt{\frac{Rg \sin \theta}{\cos \theta}}$$

6. Angular momentum is expressed as $L = Rmv \sin 90°$, because $\vec{R}$ and $\vec{p}$ are tangential to each other. Thus,

$$L = Rm\sqrt{\frac{Rg \sin \theta}{\cos \theta}} = \sqrt{\frac{m^2gR^3 \sin \theta}{\cos \theta}}$$

Because $R = l \sin \theta$, then

$$L = \sqrt{\frac{m^2gl^3 \sin^4 \theta}{\cos \theta}}$$

PROBLEM 48 Conservation of Angular Momentum

Before we begin...

1. The law of conservation of angular momentum states that in the absence of a net external torque, the angular momentum of a system will remain unchanged. The angular momentum at any time will be equal to the angular momentum at any other time.

2. Angular momentum is related to the moment of inertia of a system by

$$L = I\omega$$

3. The given information is

$$I_{\text{m-g-r}} = 250 \text{ kg·m}^2$$

$$\omega_{\text{m-g-r}} = 10 \text{ rev/min} = 1.05 \text{ rad/s}$$

$$R = 2.0 \text{ m}$$

$$m_c = 25 \text{ kg}$$

4. The equation for computing the moment of inertia of a revolving point mass particle is

$$I = mR^2$$

Solving the problem

5. The angular momentum of the merry-go-round is expressed as

$$L = I\omega = (250 \text{ kg·m}^2)(1.05 \text{ rad/s}) = 262 \text{ J·s}$$

6. After the child jumps onto the merry-go-round, the new moment of inertia of the system is

$$I = I_{\text{m-g-r}} + I_c = 250 \text{ kg·m}^2 + (25 \text{ kg})(2.0 \text{ m})^2$$
$$= 350 \text{ kg·m}^2$$

7. The new rotational speed is

$$L_1 = L_2, \quad \text{or} \quad I_1\omega_1 = I_2\omega_2,$$

$$\text{or} \quad 262 \text{ J·s} = (350 \text{ kg·m}^2)\omega_2$$

Therefore, $\omega_2 = 0.75$ rad/s.

Module 8 Simple Harmonic Motion

PROBLEM 49 Simple Harmonic Motion

Before we begin...

1. The mass of the particle does not matter because the equation of motion for the particle is independent of the mass.

2. The given information is

$$f = 3.0 \text{ Hz} \qquad A = 5.0 \text{ cm}$$

3.

Solving the problem

4. The distance between the equilibrium point and the maximum displacement is 5.00 cm, therefore the total distance traveled in three cycles is 60 cm.

5. The expression for $v(t)$ is

$$v = \frac{dy}{dt} = \omega A \cos(\omega t + \phi)$$

The maximum value of v is ωA. This occurs when $y = 0$.

6. The expression for $a(t)$ is

$$a = \frac{dv}{dt} = -\omega^2 A \sin(\omega t + \phi)$$

The maximum magnitude for the acceleration is $\omega^2 A$. This occurs when the displacement is a maximum, $y = \pm A$.

PROBLEM 50 Physical Nature of Waves

Before we begin...

1. The general expression for a traveling wave on a string starting at $y = 0$ at $(0, 0)$ is

$$y(x, t) = A \sin\left(\frac{2\pi}{\lambda}(x - vt)\right) = A \sin(kx - \omega t)$$

2. The transverse speed and acceleration are related to the position as a function of time on a traveling wave by

$$v = \frac{dy}{dt} \qquad a = \frac{d^2y}{dt^2} = \frac{dv}{dt}$$

Solving the problem

3. The derivative of displacement with respect to time at the given position and time is evaluated as

$$v = \frac{dy}{dt} = (0.12 \text{ m})(4\pi) \cos\left[\pi\left(\frac{x}{8} + 4t\right)\right]$$

$$v(1.6 \text{ m}, 0.2 \text{ s}) = -1.51 \text{ m/s}$$

4. The acceleration is

$$a = \frac{d\left((0.12 \text{ m})(4\pi) \cos\left[\pi\left(\frac{x}{8} + 4t\right)\right]\right)}{}$$

$$a(1.6 \text{ m}, 0.2 \text{ s}) = 0$$

5. Since $k = 2\pi/\lambda = \pi/8$ and $\omega = 2\pi/T$, comparison of terms yields

$$\lambda = 16 \text{ m} \qquad T = 0.5 \text{ s} \qquad v = \lambda/T = 32.0 \text{ m/s}$$

PROBLEM 51 Mathematical Nature of Waves

Before we begin...

1. The given information is

$$A = 0.2 \text{ m}$$

$$\lambda = 0.35 \text{ m}$$

$$f = 12.0 \text{ Hz}$$

$$y(0, 0) = -0.03 \text{ m}$$

2.

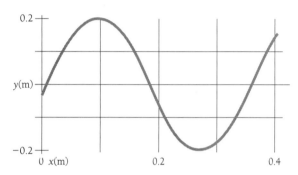

You must know the maximum and minimum values for y, the x distance before y repeats itself in phase, and the position and sign of the velocity at a given point, in order to create the plot.

3. The general expression for the wave traveling to the left as a function of position and time is given by

$$y = A \cos(kx + \omega t + \phi)$$

$$v = -\omega A \sin \omega t$$

4. The y position will be $A \cos \phi$ at $t = 0$ and $x = 0$.

(continued on next page . . .)

Solving the problem

Since the wave has a negative displacement but a positive velocity, the graph of y vs. t must begin at -0.03 m and have a positive slope at that point. It should repeat itself every 0.35 m, having a maximum displacement of ± 0.2 m.

5. Using the values of wavelength and period to identify the wave number and angular frequency, we have

$$k = 2\pi/\lambda = 18.0 \text{ rad/m} \qquad T = 1/f = 0.0833 \text{ s}$$

$$\omega = 2\pi f = 75.4 \text{ rad/s} \qquad v = f\lambda = 4.20 \text{ m/s}$$

Because the maximum displacement of the wave is ± 0.20 m and the amplitude A is 0.20 m, the phase of the wave must satisfy the boundary conditions that $y(0, 0) = -0.03$ m and that the velocity is positive at this point and time.

$$y(0, 0) = -0.03 \text{ m} = A \cos \phi$$

$$v(0, 0) = -\omega A \sin \phi$$

6. Because ω, v, and A are all greater than zero, $\sin \phi$ must be less than zero. In particular,

$$\sin \phi = -\frac{v(0, 0)}{\omega A} = -\frac{4.2}{(75.4)(0.2)} = \sin \phi$$

$$\phi = \sin^{-1}\left(-\frac{4.2}{(75.4)(0.2)}\right) = -0.2822$$

7. This particular traveling wave, using the general form and the computed values, is

$$y(x, t) = (0.2 \text{ m}) \cos [(18.0 \text{ rad/m})x + (75.4 \text{ rad/s})t + 1.42 \text{ rad}]$$

Before we begin...

1. The general wave function for a wave traveling to the left (negative x direction) as a function of position and time is

$$y(x, t) = A \cos (kx + \omega t + \phi)$$

2. The given information is

$$T = 0.025 \text{ s} \quad v_x = 30 \text{ m/s} \quad y(0, 0) = 0.02 \text{ m}$$

$$v_y(0, 0) = -2.00 \text{ m/s}$$

3. The period of a wave T is related to the angular frequency ω by the equation

$$T = 2\pi/\omega$$

4. Initial phase angle refers to the angle at $x = 0$ and $t = 0$. It defines ϕ.

Solving the problem

5. Solving for A, we use $A \cos \phi = 0.02$ m, and $A\omega \sin \phi = 2.00$ m/s, and $\omega = 80\pi \text{ s}^{-1}$ to get

$$A^2 = (0.02 \text{ m})^2 + \left(\frac{2.0 \text{ m/s}}{80\pi \text{ s}^{-1}}\right)^2$$

$$A = 0.0215 \text{ m}$$

6. Substitution back in the expressions for $y(0,0)$ yields

$$\phi = \cos^{-1} (0.02 \text{ m}/0.0215 \text{ m}) = 0.379 \text{ rad}$$

7. The maximum transverse wave speed is

$$v_{y \text{ max}} = (0.0215 \text{ m})(80\pi) = 5.41 \text{ m/s}$$

8. Evaluating k, we use $k = 2\pi/\lambda$,

$$k = 8\pi/3$$

9. The specific wave function can be expressed as

$$y(x,t) = (0.0215 \text{ m}) \cos [(8\pi/3)x + (80\pi)t + 0.379]$$

PROBLEM 53 Mathematical Nature of Waves

Before we begin...

1. The expression for the wave function of a traveling wave is

$$y(x, t) = A \cos (kx - \omega t + \phi)$$

2. The wave described in the general case is traveling in the positive x direction.

Solving the problem

3. A term-by-term comparison of the specific equation and the general equation yields

$$A = 0.25 \text{ m} \quad k = 0.30 \text{ m}^{-1} \quad \omega = 40 \text{ s}^{-1}$$

4. Using $k = 2\pi/\lambda$, we find that the wavelength is

$$\lambda = 21 \text{ m}$$

and with $\omega = 2\pi f$ that the wave speed is

$$v = (\omega/2\pi)(2\pi/k) = 130 \text{ m/s}$$

5. An inspection of the sign in front of the time component of the wave function tells us that the wave is moving to the right (in the $+x$ direction).

PROBLEM 54 Hooke's Law and the Equation of Motion

Before we begin...

1. The given variables and terms are

$$m = 1.0 \text{ kg} \quad k = 25 \text{ N/m} \quad A = 3 \text{ cm} = 0.03 \text{ m}$$

2. To solve the problem, we must (a) compute the period, (b) compute the maximum speed and maximum acceleration and (c) write equations expressing displacement, velocity, and acceleration as functions of time.

Solving the problem

3. We know that

$$\omega = \sqrt{\frac{k}{m}} = 5.0 \text{ rad/s} \quad \text{and} \quad T = \frac{2\pi}{\omega}$$

so $T = 1.26$ s.

4. Setting $^1/_2 mv^2 = ^1/_2 kA^2$ and solving for v, we find

$$v - \sqrt{\frac{k}{m}} A$$
$$= 15 \text{ cm/s} = 0.15 \text{ m/s}$$

5. Setting $ma = -kx = kA$ and solving for a, we find that

$$a = 75 \text{ cm/s}^2 = 0.75 \text{ m/s}^2$$

6. Substituting the given value of A and the computed value of ω into the equation for position as a function of time, we have to evaluate the initial conditions to find ϕ. The spring will be at $x = -3.0$ cm $= -0.03$ m at $t = 0$, so $\phi = \pi$ radians.

$$x(t) = 0.03 \cos (5t + \pi) \text{ m}$$
$$v(t) = -0.15 \sin (5t + \pi) \text{ m/s}$$
$$a(t) = -0.75 \cos (5t + \pi) \text{ m/s}^2$$

All expressions and values have been computed, and the problem is solved. It is worth pointing out that had we solved for the general expressions before solving for the maximum values of velocity and acceleration, the answers for maximum values could have been found by inspection. The maximum value of a sine or cosine function is ± 1.0. Thus the maximum velocity is 0.15 m/s and the maximum acceleration is 0.75 m/s^2.

Saunders Core Concepts in Physics Workbook

PROBLEM 55 SHM and Waves in the Real World

Before we begin...

1. For small amplitudes, the relation between the period of motion of a simple pendulum, the pendulum's length, and the acceleration due to gravity is

$$T = 2\pi\sqrt{\frac{1}{g}}$$

2. The change in a quantity as measured under two different conditions is calculated by

$$\Delta T = T_2 - T_1$$

Solving the problem

3. Calculating the periods for each of the given acceleration values, we find

$$T_2 = 2\pi\sqrt{\frac{3.00 \text{ m}}{9.79 \text{ m/s}^2}} \qquad T_1 = 2\pi\sqrt{\frac{3.00 \text{ m}}{9.80 \text{ m/s}^2}}$$

4. Subtracting the value computed for $g = 9.80$ m/s^2 from the value computed when $g = 9.79$ m/s^2,

$$\Delta T = 2\pi\sqrt{1}\left(\frac{1}{\sqrt{g_2}} - \frac{1}{\sqrt{g_1}}\right)$$

$$= 1.78 \times 10^{-3} \text{ s}$$

PROBLEM 56 Speed of a Wave in a Medium

Before we begin...

1. The given information is

$$\mu = 8.00 \text{ g/m} = 0.00800 \text{ kg/m} \quad v = 60.0 \text{ m/s}$$

2.

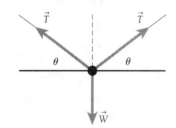

3. The tension in each section of the string is the same, ensuring that the x components of the forces equal zero, because the tensions make the same angle with respect to the x axes. The two y components together will be equal to the weight.

4. From the original drawing of the problem, we use one of the right triangles formed by the string and the span of the supports to get

$$\theta = \cos^{-1}\left(\frac{3 \text{ } L/8}{L/2}\right) = 41.4°$$

Solving the problem

5. Applying the condition of equilibrium to the y components and solving for the tension T,

$$T \sin \theta + T \sin \theta = mg$$

$$2T \sin \theta = mg$$

$$T = (mg)/(2 \sin \theta)$$

6. Using the above relationship between the tension and the weight and $T = v^2\mu$, we find

$$v^2\mu = (mg)/(2 \sin \theta)$$

$$(60 \text{ m/s})^2(0.008 \text{ kg/m}) = m(9.80 \text{ m/s}^2)/(2 \sin 41.4°)$$

$$m = 3.9 \text{ kg}$$

PROBLEM 57 Energy and Power in Waves

Before we begin...

1. The power delivered by a transverse wave is expressed by

$$P = \frac{1}{2}\mu\omega^2 A^2 v$$

2. The length of the rope does not affect the power.

Solving the problem

3. If the speed remains constant and only the length changes, the power does not change.

If A is doubled and ω is halved, the power does not change, because $P \propto \omega^2 A^2$.

Because $\omega \propto 1/\lambda$, the power does not change when both λ and A are doubled.

If the wavelength λ is halved, ω^2 will increase fourfold. The power will quadruple.

PROBLEM 58 Superposition and Interference

Before we begin...

1. The wave function of each wave is

$$y_1 = (4.0 \text{ cm}) \sin (kx - \omega t)$$

$$y_2 = (4.0 \text{ cm}) \sin (kx - \omega t - \phi)$$

2. The superposition principle states that the resultant wave function of two or more traveling waves in a medium is the sum of the wave functions of the individual waves.

$$y = y_1 + y_2$$

Solving the problem

3. Substituting into the superposition principle and using the trig identity relating $\sin (a) + \sin (b)$, we find

$$\begin{aligned} y &= (4.0 \text{ cm}) \sin (kx - \omega t) \\ &+ (4.0 \text{ cm}) \sin (kx - \omega t - 90°) \\ &= 2(4.0 \text{ cm}) \sin (kx - \omega t - 45°) \cos 45° \end{aligned}$$

If we rearrange the equation, we can write it as

$$y = [(2)(4.0 \text{ cm}) \cos 45°] \sin (kx - \omega t - 45°)$$

Inspection of the form of the equation tells us that the amplitude is

$$A = [(2)(4.0 \text{ cm}) \cos 45°] = 5.7 \text{ cm}$$

PROBLEM 59 Standing Waves

Before we begin...

1. Node positions are stationary points on the resulting wave.

2. The sum of the displacements of all the waves must result in a zero displacement in order for a point to be a node.

Solving the equation

3. Using the superposition principle to find the wave function of the combined waves, we find

$$y_1 + y_2 = A \sin (kx - \omega t) + A \sin (2kx + \omega t)$$

$$= 2A \cos\left(\frac{2kx + \omega t - kx + \omega t}{2} \right)$$

$$\times \sin \left(\frac{2kx + \omega t + kx - \omega t}{2} \right)$$

The above equation simplifies to

$$2A \cos \left(\frac{kx + 2\omega t}{2} \right) \sin \left(\frac{3kx}{2} \right)$$

4. Stationary nodes occur as a result of a term that is independent of time t equaling zero. This is true when

$$\sin \left(\frac{3kx}{2} \right) = 0$$

For this to be satisfied, $3kx = 0, \pm 2\pi, \pm 4\pi \cdots$, therefore $x = 0, 2\pi/3k, 4\pi/3k$, etc.

5. The time dependent nodes satisfy the equation

$$\cos \left(\frac{kx + 2\omega t}{2} \right) = 0$$

The solutions to this equation are

$$x = \pm (\pi - 2\omega t)/k,$$
$$\pm (3\pi - 2\omega t)/k,$$
$$\pm (5\pi - 2\omega t)/k, \text{ etc.}$$

PROBLEM 60 Standing Waves—Fixed at Both Ends

Before we begin...

1. The equation for the allowable wavelengths for a wave fixed at both ends is

$$\lambda_n = 2L/n \ (n = 1, 2, 3, ...)$$

2. The speed of the wave can be calculated for a wire under tension using

$$v = \sqrt{\frac{F_t}{\mu}}$$

Solving the problem

3. $\mu = M/L = 0.10 \text{ kg}/2.0 \text{ m} = 0.050 \text{ kg/m}$

4. The first allowable frequency of vibration is

$$f_1 = \frac{v}{\lambda_1} = 5.0 \text{ Hz}$$

For the $n = 2$ and $n = 3$ modes,

$$f_2 = 10 \text{ Hz} \quad \text{and} \quad f_3 = 15 \text{ Hz}$$

Notice that for waves fixed at both ends, the frequency $f_n = nf_1$.

Before we begin...

1. The relationship between the length of the system and the allowable frequencies for this type of system is

$$f_n = n\frac{v}{4L} \qquad (n = 1, 3, 5, ...)$$

This can be rewritten in terms of n representing the mode number of the vibration as

$$f_n = (2n - 1)\frac{v}{4L} \qquad (n = 1, 2, 3, ...)$$

2. The given information is

$$f_a = 52.0 \text{ Hz} \qquad\qquad f_b = 60.0 \text{ Hz}$$

Solving the problem

3. Writing the expressions for the nth allowable frequency for $a = n$ and $b =$ next allowable frequency, we find

$$f_a = (2n - 1)\frac{v}{4L} \qquad f_b = [2(n + 1) - 1]\frac{v}{4L}$$

4. The simultaneous equations to solve are written as

$$f_b - f_a = 60.0 \text{ Hz} - 52.0 \text{ Hz} = 8.0 \text{ Hz}$$

and

$$f_b - f_a = [2(n + 1) - 1]\frac{v}{4L} - (2n - 1)\frac{v}{4L}$$

5. After simplifying, the solution gives us the expression

$$8.0 \text{ Hz} = \frac{2v}{4L}$$

from which we find $L = 21.5$ m, the depth of the well.

PROBLEM 62 Resonance

Before we begin...

1. The given information is

$$m = w/g = 4.08 \text{ kg} \qquad k = 200 \text{ N/m}$$

$$A = 2.00 \text{ cm} \qquad \omega = 2\pi f = 62.8 \text{ s}^{-1}$$

Solving the problem

2. The second derivative of x with respect to time t is

$$\frac{d^2x}{dt^2} = -\omega_0^2 C \sin \omega_0 t - \omega^2 A \sin \omega t$$

3. Substitution into the equation of motion gives

$$-\omega_0^2 C \sin \omega_0 t - \omega^2 A \sin \omega t$$
$$= -\omega_0^2 (C \sin \omega_0 t + A \sin \omega t) + a_0 \sin \omega t$$

4. Simplifying the equation and solving for A gives

$$-\omega^2 A \sin \omega t = -\omega_0^2 A \sin \omega t + a_0 \sin \omega t$$

so that

$$A = \frac{a_0}{\omega_0^2 - \omega^2}$$

5. The maximum value of the force occurs when all of the amplitude of the motion is in A, thus $C = 0$. The amplitude of the oscillation is equal to the absolute value of A. Solving for F_0, we find

$$F_0 = |(4.08 \text{ kg})(0.0200 \text{ m})\left[\left(\frac{200 \text{ N/m}}{4.08 \text{ kg}}\right)\right.$$
$$\left. - (2\pi \times 10.0 \text{ Hz})^2\right]|$$
$$= 318 \text{ N}$$

PROBLEM 63 Basic Concepts of Thermodynamics

Before we begin...

1. Thermal equilibrium is reached when no net heat flows from one object to the other. Objects in thermal equilibrium are said to be at the same temperature.

2. The given information is

$$V_w = 500 \text{ ml} \qquad T_{1w} = 30° \text{ C}$$

$$m_{ice} = 25 \text{ g} \qquad T_{1\text{-ice}} = 0° \text{ C}$$

Solving the problem

3. The mass of the water is

$$m = (10^3 \text{ kg/m}^3)(1 \text{ m}^3/10^3 \text{ l})(0.500 \text{ l})$$
$$= 0.500 \text{ kg}$$

4. The thermal energy that would be released by lowering the water to the freezing point is

$$Q = mc\Delta T = (0.500 \text{ kg})(4186 \text{ J/kg C°})(0° \text{ C} - 30° \text{ C})$$
$$= -6.28 \times 10^4 \text{ J}$$

(The minus sign indicates that energy is released rather than absorbed.)

5. The thermal energy required to melt all of the ice is

$$Q = mL_f = (0.0250 \text{ kg})(3.33 \times 10^5 \text{ J/kg})$$
$$= 8.33 \times 10^3 \text{ J}$$

6. Because the water has more available thermal energy than that required to melt the ice, the water does not cool all the way to 0° C. Once all the ice has melted, the combined pool of water will reach equilibrium at some intermediate temperature T_2, which we can find using energy conservation

$$m_w c_w (T_2 - 30°C) + 8.33 \times 10^3 \text{ J}$$
$$+ m_{ice} c_w (T_2 - 0° \text{ C}) = 0$$

$$T_2 = 24.6° \text{ C}$$

7. In the second part of the problem, the thermal energy that would be required to melt the ice is

$$Q = mL_f = (0.250 \text{ kg})(3.33 \times 10^5 \text{ J/kg})$$
$$= 8.33 \times 10^4 \text{ J}$$

Recall that it takes only 6.28×10^4 J to cool the initial half liter of water to 0° C. Since this is not enough to melt all the ice, the system comes to equilibrium at 0° C with some of the ice still frozen. The amount of ice that is melted is calculated by applying the law of conservation of energy to the isolated system:

$$\Delta m L_f - 6.28 \times 10^4 \text{ J} = 0$$

$$\Delta m = 0.189 \text{ kg}$$

PROBLEM 64 The Ideal Gas

Before we begin...

1. The ideal gas law is expressed by the relationship

$$PV = Nk_BT$$

where P is the pressure in atmospheres, V the volume, N the number of gas molecules present, and T the temperature in Kelvins. The term k_B is a constant called Boltzmann's constant.

2. The given information is

$$T_1 = 10.0° \text{ C} \quad T_2 = 80.0° \text{ C} \quad P_1 = 2.50 \text{ atm}$$

Solving the problem

3. On the Kelvin scale, the temperatures T_1 and T_2 are

$$T_1 = 10.0° + 273° = 283 \text{ K}$$

and

$$T_2 = 80.0° + 273° = 353 \text{ K}$$

4. Collecting the relevant quantities P and T on one side of the ideal gas law,

$$\frac{P}{T} = \frac{Nk_B}{V}$$

Because V and N are constant in this problem, we can write

$$\frac{P_1}{T_1} = \frac{P_2}{T_2}$$

Solving for P_2 we find

$$P_2 = \frac{P_1 T_2}{T_1} = \frac{(2.50 \text{ atm})(353 \text{ K})}{(283 \text{ K})}$$
$$= 3.12 \text{ atm}$$

PROBLEM 65 The First Law of Thermodynamics

Before we begin...

1. The work done by a gas in expanding through a volume is expressed by

$$dW = PdV$$

therefore

$$W = \int_{V_1}^{V_2} PdV$$

2. The initial and final volumes of this gas are

$$V_1 = 12 \text{ m}^3 \qquad V_2 = 36 \text{ m}^3$$

Solving the problem

3. Using the definition of work, we have

$$W = \int_{V_1}^{V_2} (P_0)\, e^{-bV} dV = \frac{-P_0}{b} (e^{-bV_2} - e^{-bV_1})$$
$$= -(12 \text{ atm})(12 \text{ m}^3)(e^{-3} - e^{-1})$$

4. Using the conversion between atmospheres and N/m^2 to give proper SI units for our answer, we find

$$12 \text{ atm} = (12 \text{ atm})(1.013 \times 10^5 \text{ N/m}^2/1 \text{ atm})$$
$$= 12.2 \times 10^5 \text{ N/m}^2$$

$$W = -(12.2 \times 10^5 \text{ N/m}^2)(12 \text{ m}^3)(e^{-3} - e^{-1})$$
$$= 4.64 \times 10^6 \text{ J}$$

PROBLEM 66 Carnot Engines

Before we begin...

1. The given information is

$$T_c = 20° \text{ C} \qquad T_h = 500° \text{ C} \qquad P = 150 \text{ kW}$$

2. Power P is the rate at which work is done.

$$P = \frac{\Delta W}{\Delta t}$$

3. The efficiency of any heat engine is expressed by the relationship

$$e = \frac{W}{Q_h}$$

where W is the work done by the engine and Q_h is the heat absorbed from its hot reservoir.

4. For a Carnot engine, the efficiency is given by

$$e_C = \frac{(T_h - T_c)}{T_h}$$

where all temperatures are in kelvins.

Solving the problem

5. Solving for the efficiency of the engine we find

$$e_C = \frac{(T_h - T_c)}{T_h} = \frac{480 \text{ K}}{773 \text{ K}} = 0.621$$

6. The rate at which work is being done by the engine is

$$1.50 \times 10^5 \text{ W}$$

so in one hour (3600 s) the work will be

$$W = (1.50 \times 10^5 \text{ W})(3600 \text{ s}) = 5.40 \times 10^8 \text{ J}$$

7. The heat absorbed will be

$$Q_h = W/e_C = 8.70 \times 10^8 \text{ J}$$

8. Because $W = Q_h - Q_c$, we compute

$$Q_c = 8.70 \times 10^8 \text{ J} - 5.40 \times 10^8 \text{ J} = 3.30 \times 10^8 \text{ J}$$

PROBLEM 67 Carnot Engines—The Heat Pump

Before we begin...

1. The Carnot COP is computed by the relationship

$$\text{COP} = \frac{Q_h}{W} = \frac{Q_h}{Q_h - Q_c} = \frac{T_h}{T_h - T_c}$$

2. The given information is

$$T_h = 22° \text{ C} \qquad\qquad T_c = -3° \text{ C}$$

Solving the problem

3. The temperatures T_h and T_c on the Kelvin scale are

$$T_h = 22° \text{ C} = 295 \text{ K} \quad \text{and} \quad T_c = -3° \text{ C} = 270 \text{ K}$$

4. We can evaluate the COP as

$$\frac{T_h}{T_h - T_c} = \frac{295 \text{ K}}{25 \text{ K}} = 11.8$$

Solutions

PROBLEM 68 Applications of Entropy— General Process for an Ideal Gas

Before we begin...

1. The given information is

$$T_i = 25° \text{ C} = 298 \text{ K}$$

$$T_f = -18° \text{ C} = 255 \text{ K}$$

$$P_i = 1.00 \text{ atm}$$

Solving the problem

2. In the constant volume part of the problem, computing the change in entropy ΔS while using $C_V = 5R/2$ for an ideal gas, we get

$$\Delta S = nC_V \ln \frac{T_f}{T_i} + nR \ln \frac{V_f}{V_i} = n\frac{5R}{2} \ln \left(\frac{255 \text{ K}}{298 \text{ K}}\right) + 0$$

$$= n(5R/2)(-0.156)$$

$$= -0.390 \ nR$$

3. Because the volume does change in the second part of the problem, we use

$$V = nRT/P$$

to give $V_f/V_i = T_f/T_i$. Substituting into the general relationship for change in entropy gives us the relationship

$$\Delta S = n(5R/2) \ln \frac{T_f}{T_i} + nR \ln \frac{T_f}{T_i} = n(7R/2) \ln \frac{T_f}{T_i}$$

Using the given temperatures, we find that the change in entropy is

$$\Delta S = n(7R/2)(-0.156) = -0.545 \ nR$$

PROBLEM 69 Applications of Entropy

Before we begin...

1. When a small amount of heat dQ flows into a system at temperature T (measured on an absolute scale such as kelvins), the change in the system's entropy is

$$dS = dQ/T$$

2. The iceberg merges with the seawater in a three-stage process: (a) the ice warms from its initial temperature to its melting point; (b) the ice melts at constant temperature; and, (c) the water that was once an iceberg warms from the melting point to the temperature of the surrounding seawater. In each of these stages, heat flows from the seawater to the iceberg (or former iceberg).

3. The given information is

$$m = 1.0 \times 10^3 \text{ kg} \quad T_i = 268 \text{ K} \quad T_f = 278 \text{ K}$$

Solving the problem

4. As the ice changes temperature by a small amount dT, it absorbs heat equal to

$$dQ = mc_{\text{ice}}dT$$

Thus the total entropy gain of the ice as it warms from $T_i = 268$ K to $T_{\text{m-p}} = 273$ K is

$$\Delta S = \int \frac{dQ}{T} = \int mc_{\text{ice}} \frac{dT}{T} = mc_{\text{ice}} \ln \left(\frac{T_{\text{m-p}}}{T_i}\right)$$

$$=$$

$$(1.00 \times 10^5 \text{ kg})(2090 \text{ J/kg K})$$
$$\ln (273/268) = 3.86 \times 10^6 \text{ J/K}$$

5. As the ice melts, it absorbs heat $\Delta Q = mL_f = 3.33 \times 10^{10}$ J. Since this takes place at constant temperature $T_{\text{m-p}} = 273$ K, the change in the iceberg's entropy is

$$\Delta S = \frac{\Delta Q}{T_{\text{m-p}}} = 1.22 \times 10^8 \text{ J/K}$$

(continued on next page . . .)

Saunders Core Concepts in Physics Workbook

6. As the newly melted iceberg warms from $T_{\text{m-p}} = 273$ K to $T_f = 278$ K, the change in its entropy is

$$\Delta S = \int \frac{dQ}{T} = \int mc_{\text{water}} \frac{dT}{T} = mc_{\text{water}} \ln\left(\frac{T_{\text{m-p}}}{T_i}\right)$$

$$= (1.00 \times 10^5 \text{ kg})(4186 \text{ J/kg K}) \ln(278/273)$$

$$= 7.60 \times 10^6 \text{ J/K}$$

7. Throughout these processes, the total amount of heat that the ocean gives up is equal to that absorbed by the iceberg

$$\Delta Q = -mc_{\text{ice}}(T_{\text{m-p}} - T_i) - mL_f$$
$$\quad - mc_{\text{water}}(T_{\text{m-p}} - T_i)$$
$$= -3.644 \times 10^{10} \text{ J}$$

As it gives off this heat, the seawater remains at a constant temperature $T_f = 278$ K. (The ocean is presumably so big that it does not cool noticeably as the iceberg melts.) Thus, the entropy lost by the seawater during these processes is

$$\Delta S = \frac{\Delta Q}{T_f} = -1.288 \times 10^8 \text{ J/K}$$

8. Adding together the three stages of entropy gained by the iceberg and the entropy lost by the surrounding seawater yields the total change in entropy

$$\Delta S_{\text{tot}} = 3.86 \times 10^6 \text{ J/K} + 1.22 \times 10^8 \text{ J/K}$$
$$\quad + 7.60 \times 10^6 \text{ J/K} - 1.288 \times 10^8 \text{ J/K}$$
$$= 4.66 \times 10^6 \text{ J/K}$$

This is an archetypical example of the second law of thermodynamics at work: some parts of the universe may lose entropy as other parts gain it, but the net change is always an increase in the total entropy of the universe.

PROBLEM 70 Coulomb's Law

Before we begin...

1. The given information is

$$q = +10.0 \ \mu C \qquad L = 60.0 \text{ cm} = 0.600 \text{ m}$$

$$W = 15.0 \text{ cm} = 0.150 \text{ m}$$

2. Coulomb's law for the force between two point charges states that the two charges will experience a force that is described by the relationship

$$\vec{F}_{21} = k_e \frac{q_1 q_2}{r^2} \hat{r}_{12}$$

where k_e is a constant and r is the distance between the charges. The force is along the line adjoining the two charges.

3. The direction of the force is attractive when the charges are of opposite sign. If the charges are of like sign, the force is repulsive.

4.

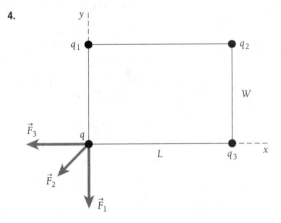

(continued on next page . . .)

Solving the problem

5. Using Coulomb's law, we can evaluate the forces $\vec{F}_1$, $\vec{F}_2$, and $\vec{F}_3$ as

$$F_1 = k_e \frac{q_1 q_2}{r_1^2} = (9.00 \times 10^9 \text{ N·m}^2/\text{C}^2)$$
$$\times \frac{(1.00 \times 10^{-5} \text{ C})(1.00 \times 10^{-5} \text{ C})}{(0.150 \text{ m})^2}$$
$$= 40.0 \text{ N}$$

The direction of $\vec{F}_1$ is 270° with respect to the x direction because the charges repel each other.

$$F_2 = \frac{k_e q_2 q}{r_2^2} = (9.00 \times 10^9 \text{ N·m}^2/\text{C}^2)$$
$$\times \frac{(1.00 \times 10^{-5} \text{ C})(1.00 \times 10^{-5} \text{ C})}{(0.600 \text{ m})^2 + (0.150 \text{ m})^2}$$
$$= 2.35 \text{ N}$$

To find the direction, we evaluate

$$\theta = \tan^{-1} \frac{0.150 \text{ m}}{0.600 \text{ m}} = 14°$$

The direction of $\vec{F}_2$ is $180° + 14° = 194°$ with respect to the positive x axis because the charges repel each other.

$$F_3 = k_e \frac{q_3 q}{r_3^2} = (9.00 \times 10^9 \text{ N·m}^2/\text{C}^2)$$
$$\times \frac{(1.00 \times 10^{-5} \text{ C})(1.00 \times 10^{-5} \text{ C})}{(0.600 \text{ m})^2}$$
$$= 2.50 \text{ N}$$

The direction of $\vec{F}_3$ is 180°.

6. Resolving the forces into their x and y components, we find

$$F_{1x} = 0 \qquad\qquad F_{1y} = -40.0 \text{ N}$$
$$F_{2x} = -2.28 \text{ N} \qquad F_{2y} = -0.569 \text{ N}$$
$$F_{3x} = -2.50 \text{ N} \qquad F_{3y} = 0$$

so that

$$F_{Tx} = -4.78 \text{ N} \qquad F_{Ty} = -40.6 \text{ N}$$

7. The magnitude of $\vec{F}_T$ is

$$F_T = \sqrt{F_x^2 + F_y^2}$$
$$= \sqrt{(-4.78)^2 + (-40.6)^2} = 40.9 \text{ N}$$

8. The direction of $\vec{F}_T$ is

$$\theta = \tan^{-1} \frac{F_y}{F_x} = \tan^{-1} \frac{-40.7}{-4.78} = 263°$$

Before we begin...

1.

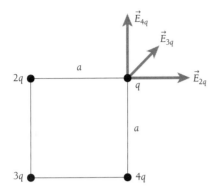

2. The electric field for a point charge is calculated using the relationship

$$\vec{E} = \frac{\vec{F}_e}{q_o} = k_e \frac{q}{r^2} \hat{r}$$

3. The direction of the electric field due to a point charge points radially away from a positive charge creating the field and radially toward a negative charge creating the field.

4. The electric force on a charge q is given by $\vec{F} = q\vec{E}$.

Solving the problem

5. Evaluating the fields $\vec{E}_{2q}$, $\vec{E}_{3q}$ and $\vec{E}_{4q}$, we find

$$E_{2q} = k_e(2q)/a^2$$

(The direction of $\vec{E}_{2q}$ is 0° or along the +x axis because the charge is positive, thus the field points away from the charge.)

$$E_{3q} = k_e(3q)/(\sqrt{2}a)^2$$

(The direction of $\vec{E}_{3q}$ is 45° because the charge is positive.)

$$E_{4q} = k_e(4q)/a^2$$

(The direction of $\vec{E}_{4q}$ is 90° because the charge is positive.)

6. Resolving the fields into their x and y components we find

$$E_{2qx} = k_e(2q)/a^2 \quad E_{2qy} = 0$$

$$E_{3qx} = k_e(3q)(\cos 45°)/(\sqrt{2}a)^2$$

$$E_{3qy} = k_e(3q)(\sin 45°)/(\sqrt{2}a)^2$$

$$E_{4qx} = 0 \qquad E_{4qy} = k_e(4q)/a^2$$

therefore

$$E_{Tx} = (3.06\ k_e)q/a^2 \quad \text{and}$$
$$E_{Ty} = (5.06\ k_e)q/a^2$$

7. Calculating the magnitude of $\vec{E}_T$, we find

$$E_T = \sqrt{E_x^2 + E_y^2}$$
$$= \frac{kq}{a^2}\sqrt{(3.06)^2 + (5.06)^2} = 5.91\ \frac{kq}{a^2}$$

8. The direction is determined to be

$$\theta = \tan^{-1}\frac{E_y}{E_x} = 58.8°$$

9. The electric force $\vec{F}$ points in the same direction as $\vec{E}_T$ (if q is positive) and has magnitude

$$F = qE_T = 5.91\ \frac{kq}{a^2}$$

PROBLEM 72 Gauss's Law—Electric Flux

Before we begin...

1. The electric flux is calculated by the relationship

$$\Phi = \vec{E} \cdot \vec{A} = EA \cos \theta$$

where θ is the angle between $\vec{E}$ and $\vec{A}$.

2. The given information is

$$\vec{E} = (a\hat{i} + b\hat{j})$$

Solving the problem

3. The surface vectors $\vec{A}_1$, $\vec{A}_2$, and $\vec{A}_3$ that correspond to the three questions in the description are

$$\vec{A}_1 = A\hat{i} \qquad \vec{A}_2 = A\hat{j} \qquad \vec{A}_3 = A\hat{k}$$

4. Evaluating the flux for each of the questions, we determine that

$$\Phi_1 = (a\hat{i} + b\hat{j}) \cdot A\hat{i} = aA$$

$$\Phi_2 = (a\hat{i} + b\hat{j}) \cdot A\hat{j} = bA$$

$$\Phi_3 = (a\hat{i} + b\hat{j}) \cdot A\hat{k} = 0$$

PROBLEM 73 Gauss's Law

Before we begin...

1. Gauss's law for electric fields states that the total electric flux is equal to the net enclosed charge divided by a constant.

$$\Phi = q_{in}/\epsilon_0$$

2. Because the electric flux does not depend upon the shape of the closed surface, the charge can be calculated from the flux without regard to the type of surface. It is only when we are evaluating the electric field $\vec{E}$ that the type of surface is selected for symmetry.

3. If the electric flux is negative, $\vec{E}$ and $\vec{A}$ must point in opposite directions. Under this condition, the net charge within the surface must be negative.

Solving the problem

4. Using the relationship between total electric flux and the net charge, we find

$$q_{in} = \epsilon_0 \Phi = (8.85 \times 10^{-12} \ C^2/N \cdot m^2)$$
$$\times (8.60 \times 10^4 \ N \cdot m^2/C)$$
$$= 7.61 \times 10^{-7} \ C$$

5. Because the flux is positive, the net charge within the surface must also be positive.

If the flux had the same magnitude but were negative, the magnitude of the charge enclosed by the surface would not change. The sign of the charge, however, would be negative.

Before we begin...

1. The charge inside a right cylinder of length l and radius r would be

$$q_{in} = \lambda l$$

2. The flux through the two end caps of the cylinder will equal zero because $\vec{E}$ and $d\vec{A}$ are perpendicular.

Solving the problem

3. Applying Gauss's law to derive the electric field at radial distances from a long straight wire, we have three surfaces over which to integrate: the two end caps and the side surface surrounding the filament. The end caps yield zero flux.

Evaluating the flux through the side surface, we have

$$\Phi = \oint \vec{E} \cdot d\vec{A} = E \oint dA = E(2\pi rl)$$

because $\vec{E}$ on the cylindrical surface is constant and parallel to $d\vec{A}$.

By setting the flux $\Phi = q_{in}/\epsilon_0$ and using $q_{in} = \lambda l$, we have

$$E(2\pi rl) = q_{in}/\epsilon_0 = \lambda l/\epsilon_0$$

Therefore,

$$E = \lambda/2\pi\epsilon_0 r$$

4. In this problem, λ is negative, indicating the direction of $\vec{E}$ is toward the filament rather than away from it. Evaluating for $r_1 = 10.0$ cm $= 0.1$ m, we find

$$E_1 = (-90.0 \times 10^{-6} C/m)$$
$$/[(2\pi)(8.85 \times 10^{-12} C^2/N\cdot m^2)(0.100 \text{ m})]$$

$$= -1.62 \times 10^7 \text{ N/C}$$

Solving for E_2 and E_3, we find

$$E_2 = -8.10 \times 10^6 \text{ N/C}$$
$$E_3 = -1.62 \times 10^6 \text{ N/C}$$

Before we begin...

1. The electric potential difference is related to the change in electric potential energy by

$$\Delta V = \frac{\Delta U}{q}$$

so, for point charges letting $V_\infty = 0$, we find

$$V = k_e \frac{q}{r}$$

2. The given information is

$$q_1 = +5.00 \text{ nC} \quad q_2 = -3.00 \text{ nC} \quad r = 0.35 \text{ m}$$

Solving the problem

3. Using the given information to compute the potential energy, we have

$$U = k_e \frac{q q_0}{r} = (9 \times 10^9 \text{ N}\cdot m^2/C^2)$$

$$\times \frac{(5 \times 10^{-9} C)(-3 \times 10^{-9} C)}{(0.350 \text{ m})}$$

$$= -3.86 \times 10^{-7} \text{ J}$$

Since the potential energy is negative, work will have to be done to separate the two charges.

4. Using the definition of electric potential, we sum the potentials due to the two point charges to give

$$V = \sum V = \sum k_e \frac{q}{r} = k_e \frac{q_1}{r_1} + k_e \frac{q_2}{r_2}$$

$$= (9.00 \times 10^9 \text{ N}\cdot m^2/C^2)[(5 \times 10^{-9} C)$$
$$/(0.175 \text{ m}) + (-3 \times 10^{-9} C)/(0.175 \text{ m})]$$

$$= 103 \text{ V}$$

Solutions

PROBLEM 76 The Electric Field and Electric Potential

Before we begin...

1. The electric field is related to the electric potential in the (x, y, z) coordinate system by the relationships

$$E_x = -\frac{\partial V}{\partial x}\hat{i} \qquad E_y = -\frac{\partial V}{\partial y}\hat{j} \qquad E_z = -\frac{\partial V}{\partial z}\hat{k}$$

2. The given information is

$$V = 4xz - 5y + 3z^2$$

$$\vec{r} = (x, y, z) = (2, -1, 3) \text{ m}$$

Solving the problem

3. Computing the components of $\vec{E}$ using the relationship between $\vec{E}$ in volts/meter and V in volts, we find

$$E_x = -\frac{\partial V}{\partial x}\hat{i} = -4z\hat{i}$$

$$E_y = -\frac{\partial V}{\partial y}\hat{j} = 5\hat{j}$$

$$E_z = -\frac{\partial V}{\partial z}\hat{k} = (-4x - 6z)\hat{k}$$

4. Evaluating the components of $\vec{E}$ at $\vec{r} = (x, y, z) = (2, -1, 3)$ m, we find

$$E_x = (-12 \text{ V/m})\hat{i}$$

$$E_y = (5 \text{ V/m})\hat{j}$$

$$E_z = (-26 \text{ V/m})\hat{k}$$

PROBLEM 77 The Electric Field and Electric Potential

Before we begin...

1. A constant electric potential over a region of space tells us that the electric field over that region of space is zero. The partial derivative of a constant with respect to position is by definition equal to zero.

Solving the problem

2. Because the electric potential is constant inside the sphere, we know

$$E_x = 0 \qquad E_y = 0 \qquad E_z = 0$$

3. For the region outside of the sphere, we compute the components of the electric field to be

$$E_x = -\frac{\partial V}{\partial x}\hat{i} = -[0 + 0 + E_0a^3z(-3/2)$$

$$(x^2 + y^2 + z^2)^{-5/2}(2x)]\hat{i}$$

$$= 3E_0a^3\frac{xz}{(x^2 + y^2 + z^2)^{5/2}}\hat{i}$$

$$E_y = -\frac{\partial V}{\partial y}\hat{j} = -[0 + 0 + E_0a^3z(-3/2)$$

$$\times (x^2 + y^2 + z^2)(-5/2y)]\hat{j}$$

$$= 3E_0a^3\frac{yz}{(x^2 + y^2 + z^2)}\hat{j}$$

$$E_z = -\frac{\partial V}{\partial z}\hat{k} = -[0 - E_0 + E_0a^3z(-3/2)$$

$$\times (x^2 + y^2 + z^2)^{-5/2}(2z)$$

$$+ E_0a^3(x^2 + y^2 + z^2)^{-3/2}]\hat{k}$$

$$= \left(E_0 + E_0a^3\frac{(2z^2 - x^2 - y^2)}{(x^2 + y^2 + z^2)^{5/2}}\right)\hat{k}$$

PROBLEM 78 Magnetic Force on a
Moving Charge

Before we begin...

1. The magnetic force $\vec{F}_B$ is related to the magnetic field $\vec{B}$ and the velocity $\vec{v}$ of a moving charged particle by the cross product equation

$$\vec{F}_B = q\vec{v} \times \vec{B}$$

2. This force does not change the speed of the particle because it acts in a perpendicular direction to the velocity; therefore, it causes a centripetal acceleration.

3.

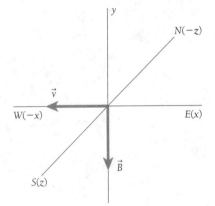

4. The given information is

$\vec{B} = -0.5 \times 10^{-4}\ T\hat{j}$ $\qquad$ $\vec{v} = (-6.2 \times 10^6\ m/s)\hat{i}$

$q = +1.60 \times 10^{-19}\ C$ $\qquad$ $m = 1.67 \times 10^{-27}\ kg$

Solving the problem

5. $\vec{F}_B = q\vec{v} \times \vec{B} = qvB \sin \theta$, where θ is the angle between $\vec{v}$ and $\vec{B}$. The direction of the initial force is the direction $(-\hat{i} \times -\hat{j}) = +\hat{k}$. In the diagram, this corresponds to southward.

$$\vec{F}_B = [(+\ 1.60 \times 10^{-19}\ C)$$
$$(6.2 \times 10^6\ m/s)(0.5 \times 10^{-4}\ T) \sin 90°]\hat{k}$$
$$= 5.0 \times 10^{-17}\ N\hat{k}$$

6. The centripetal force is related to mass m, speed v, and the radius r of the path by the equation

$$F_c = \frac{mv^2}{r}$$

7. Setting the centripetal force equal to the magnetic force allows us to solve for the radius of the path r:

$$r = \frac{mv^2}{F_c}$$
$$= \frac{(1.67 \times 10^{-27}\ kg)(6.2 \times 10^6\ m/s)^2}{5.0 \times 10^{-17}\ N}$$
$$= 1.3 \times 10^3\ m$$

Solutions

Before we begin...

1. The net force acting upon the charged particle is equal to the sum of its electrical force $\vec{F}_E$ and its magnetic force $\vec{F}_B$.

$$\vec{F} = q(\vec{E} + \vec{v} \times \vec{B})$$

2. The acceleration $\vec{a}$ is equal to the net force $\vec{F}$ divided by the mass m.

3. The given information is

$$\vec{E} = (2.5\hat{i} + 5.0\hat{j}) \text{ V/m} \quad \vec{B} = 0.40\hat{k} \text{ T}$$
$$\vec{v} = 10\hat{i} \text{ m/s} \qquad e = -1.60 \times 10^{-19} \text{ C}$$
$$m_e = 9.11 \times 10^{-31} \text{ kg}$$

Solving the problem

4. Using $\vec{F}_E = q\vec{E}$ to evaluate the electric force, we find

$$\vec{F}_E = (-1.60 \times 10^{-19} \text{ C})[(2.5\hat{i} + 5.0\hat{j}) \text{ V/m}]$$
$$= (-4.0 \times 10^{-19}\hat{i} - 8.0 \times 10^{-19}\hat{j}) \text{ N}$$

5. The magnetic force $\vec{F}_B$ can be evaluated as follows:

$$\vec{F}_B = q\vec{v} \times \vec{B}$$
$$= (-1.60 \times 10^{-19} \text{ C})(10\hat{i} \text{ m/s} \times 0.40\hat{k} \text{ T})$$
$$= (-1.60 \times 10^{-19} \text{ C})(-4.0\hat{j} \text{ m·T/s})$$
$$= 6.4 \times 10^{-19}\hat{j} \text{ N}$$

6. The Lorentz force is the sum of the two forces

$$\vec{F} = \vec{F}_E + \vec{F}_B$$
$$= (-4.0 \times 10^{-19}\hat{i} - 1.6 \times 10^{-19}\hat{j}) \text{ N}$$

Using Newton's second law, the acceleration of the electron is calculated as

$$\vec{a} = \vec{F}/m = [(-4.0 \times 10^{-19}\hat{i} - 1.6 \times 10^{-19}\hat{j}) \text{ N}]$$
$$/(9.11 \times 10^{-31} \text{ kg})$$
$$= (-4.4 \times 10^{11}\hat{i} - 1.8 \times 10^{11}\hat{j}) \text{ m/s}^2$$

Before we begin...

1. The magnetic force per unit length on a long, straight wire carrying a current $\vec{I}$ in a magnetic field $\vec{B}$ is

$$\vec{F}/L = \vec{I} \times \vec{B}$$

We know that the wire carries a current $\vec{I} = 2.0$ amps into the page, so we just need to find the magnetic field $\vec{B}$ at point A, in order to calculate the force per unit length.

2.

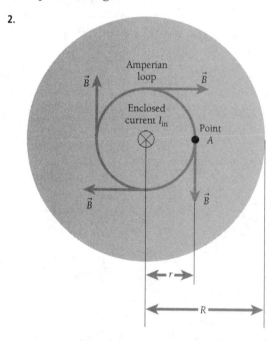

The amperian loop shown in the diagram passes through point A and shares the circular symmetry of the bundle of wires, so $\vec{B}$ should have the same magnitude everywhere around the loop. By the right-hand rule, the direction of $\vec{B}$ should be clockwise around the loop, as shown.

(continued on next page . . .)

3. Not all 100 wires thread the amperian loop in the diagram. The loop is a circle of radius $r = 0.20$ cm, while the bundle is a circle of radius $R = 0.50$ cm. Since the area of a circle is proportional to its radius squared, the loop encloses a fraction of $(r^2/R^2) = 0.16$ of the bundle. Thus 16 of the 100 wires pass through the amperian loop, carrying a current of

$$\vec{I}_{in} = 16(2.0 \text{ amps}) = 32.0 \text{ amps}$$

4. By Ampère's law, the line integral of the magnetic field around the amperian loop is μ_0 times the current $\vec{I}_{in}$ passing through it

$$\int \vec{B} \cdot d\vec{s} = \mu 0 I_{in}$$

Solving the problem

5. Since the magnetic field has a constant magnitude everywhere on the amperian loop, and is everywhere tangent to this loop, the line integral $\int \vec{B} \cdot d\vec{s}$ reduces to $B(2\pi r)$, and Ampère's law yields

$$B = \frac{\mu_0 I_{in}}{2\pi r} = 3.20 \times 10^{-3} \text{ T}$$

The wire carries a current $I = 2.0$ amps, directed into the page, through a field of strength $B = 3.20 \times 10^{-3}$ T, directed clockwise around the center of the bundle (and thus pointing "down" at point A).

6. The force per unit length on the wire is thus

$$\vec{F}/L = \vec{I} \times \vec{B} = 6.40 \times 10^{-3} \text{ N/m}$$

directed toward the center of the bundle. This force tends to hold the wires of the bundle together, in other words, although it is probably not strong enough to make much difference if other forces were attempting to pull the bundle apart.

PROBLEM 81 Magnetic Flux and Gauss's Law for Magnetism

Before we begin...

1. The magnetic flux is calculated by the relationship

$$\Phi = BA \cos \theta$$

for a flat surface with a constant magnetic field passing through it.

2.

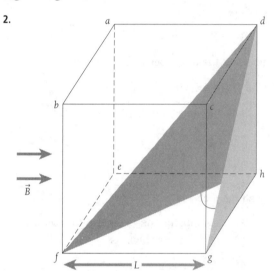

Solving the problem

3. The area of the projection of *dfhd* onto *dcgh* is $\frac{1}{2}L^2$; therefore the magnetic flux is

$$\Phi = BA \cos \theta = B(\tfrac{1}{2}L^2) = \tfrac{1}{2}BL^2$$

4. Because projecting *acfa* onto *abfe* also gives a projected area of $\frac{1}{2}L^2$, the magnetic flux in the second part of the description is also

$$\Phi = BA \cos \theta = B(\tfrac{1}{2}L^2) = \tfrac{1}{2}BL^2$$

PROBLEM 82 Faraday's Law of Induction and Lenz's Law

Before we begin...

1. Faraday's law of induction states that an emf is induced as a result of a changing magnetic flux.

$$\text{Emf} = \varepsilon = -\frac{d\Phi_B}{dt}$$

2. The total flux is calculated at any given instant by the relationship

$$\Phi_B = NBA \cos \theta$$

where θ is the angle between $\vec{B}$ and $\vec{A}$.

3. The given information is

$$A = 14 \text{ cm}^2 = 1.4 \times 10^{-3} \text{ m}^2 \qquad N = 2$$

$$B = (0.50 \text{ T})e^{-t/7} \qquad\qquad \theta = 0°$$

$$B_0 = 0.50 \text{ T}$$

Solving the problem

4. The expression for the magnetic flux as a function of time for this problem is

$$\Phi_B = NBA \cos \theta = NBA_0 e^{-t/t_0}$$

Applying Faraday's law to the changing flux to evaluate the emf we have

$$|\varepsilon| = \left| \frac{d\Phi_B}{dt} \right| = NAB_0 \left| \frac{d}{dt}(e^{-t/t_0}) \right|$$

so

$$|\varepsilon| = \frac{NAB_0}{t_0} e^{-t/t_0} = (2.0 \times 10^{-4} \text{ V})e^{-t/(7 \text{ s})}$$

PROBLEM 83 Faraday's Law of Induction and Lenz's Law

Before we begin...

1. The given information is

$$B = 3.3 \times 10^{-5} \text{ T} \quad L = 1.0 \text{ m} \quad \omega = 5\pi \text{ rad/s}$$

Solving the problem

2. Substituting into the equation for the emf of a rotating bar and evaluating, we find

$$\begin{aligned} \varepsilon &= \tfrac{1}{2}B\omega L^2 \\ &= (0.5)(3.3 \times 10^{-5} \text{ T})(5\pi \text{ rad/s})(1.0 \text{ m})^2 \\ &= 2.6 \times 10^{-4} \text{ V} \end{aligned}$$

Before we begin...

1. The general form of Faraday's law states that when a changing magnetic flux creates an emf, it must also produce an electric field

$$\oint \vec{E} \cdot d\vec{S} = -\frac{d\Phi_B}{dt}$$

2. The force acting upon a charged particle by an electric field is equal to the electric field multiplied by the charge

$$\vec{F} = q\vec{E}$$

3. The given information is

$$B = (2.0t^3 - 4.0t^2 + 0.80) \text{ T}$$

$$R = 0.025 \text{ m}$$

$$r_2 = 0.05 \text{ m}$$

Solving the problem

4. Only the region within the smaller circle (with radius R) contributes to the magnetic flux

$$\Phi_B = BA = (\pi R^2)(2.0t^3 - 4.0t^2 + 0.80) \text{ T}$$

5. Taking the derivative of the magnetic flux with respect to time to evaluate the emf as a function of time, we find

$$\varepsilon = \frac{d\Phi_B}{dt} = (\pi R^2)(6.0t^2 - 8.0t) \text{ T/s}$$

6. Setting the evaluated integral of E over the closed path equal to the rate of change of magnetic flux, we find

$$E(2\pi r_2) = (\pi R^2)(6.0t^2 - 8.0t) \text{ T/s}$$

7. Solving for E and evaluating at the given time $t = 2.0$ s, we have

$$E = (\pi R^2)(6.0t^2 - 8.0t)/(2\pi r_2)$$
$$= (2.0 \times 10^{-3} \text{ m}^2)(8.0)(\text{T/s})/(0.314 \text{ m})$$
$$= 0.050 \text{ N/C}$$

8. Using Lenz's law, the induced emf is in a direction to oppose the rate of change of magnetic flux. Because the flux is increasing into the plane, the emf must be directed counterclockwise around the circle. As a result of this, the electric field must also be counterclockwise. An electron is negative, so it will experience a clockwise force.

Solving for the force that results from the magnetic field,

$$F = qE = (1.6 \times 10^{-19} \text{ C})(0.050 \text{ N/C})$$
$$= 8.0 \times 10^{-21} \text{ N}$$

9. The time at which the electric field is zero is computed by

$$E = (\pi R^2)(6.0t^2 - 8.0t)/(2\pi r_2) = 0$$

Therefore

$$(6.0t^2 - 8.0t) = 0$$

This occurs at $t = 1.33$ s.

PROBLEM 85 Voltage, Resistance, and
Ohm's Law

Before we begin...

1. The power delivered to electric circuits is computed by using

$$P = IV$$

where I represents the electric current and V is the potential difference.

2. Identify the given information:

$V = 12.0$ volts   cost/kW·h = $0.06

rating = 55.0 A·h

Solving the problem

3. Expressing power as current times potential difference, the expression for the energy delivered in terms of current, voltage, and time t is written as

$$U = Pt = IVt$$

4. The dimensions of the equation are

[current] × [potential difference] × [time]

5. We have to multiply the rating of the battery [current] × [time] by [potential difference].

6. Rearranging the expression for energy, we determine that

$$U = IVt = (It)V$$

$$= (55.0 \text{ A·h})(12.0 \text{ V}) = 660 \text{ W·h}$$

7. To calculate the cost of this energy, we convert the W·h to kW·h by

$$660 \text{ W·h} = (660 \text{ W·h})(1 \text{ kW}/1000 \text{ W})$$

$$= 0.660 \text{ kW·h}$$

The cost of the total energy is computed to be

$$\text{cost} = (0.660 \text{ kW·h})(\$0.06/\text{kW·h})$$

$$= \$0.0396$$

The cost of the energy is certainly not what dictated the cost of the battery!

Circuit Analysis and
Kirchhoff's Laws

Before we begin...

1. For resistors in simple series, the equivalent resistance is

$$R_S = R_1 + R_2 + ...$$

2. For resistors in simple parallel, the equivalent resistance is

$$\frac{1}{R_P} = \frac{1}{R_1} + \frac{1}{R_2} + ...$$

Solving the problem

3. The resistors R and $5.0\ \Omega$ are connected in series.

4. Their equivalent resistance is $R_1 = R + 5.0\ \Omega$.

5.

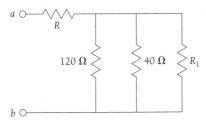

6. In the redrawn diagram, we observe that R_1, the $120\ \Omega$ resistor, and the $40\ \Omega$ resistor are connected in parallel. The equivalent resistance of these elements is found by

$$\frac{1}{R_2} = \frac{1}{R_1} + \frac{1}{40\ \Omega} + \frac{1}{120\ \Omega}$$

7.

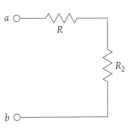

8. In this last diagram, R_2 and the resistor R are connected in series.

$$R_E = R + R_2$$

9. Substituting the values for R_2 and subsequently R_1 into the equation, we find

$$R_E = R + \frac{1}{\dfrac{1}{R_1} + \dfrac{1}{40\ \Omega} + \dfrac{1}{120\ \Omega}}$$

so

$$75\ \Omega = R + \frac{1}{\dfrac{1}{(R + 5.0\ \Omega)} + \dfrac{1}{40\ \Omega} + \dfrac{1}{120\ \Omega}}$$

Simplifying the equation

$$75\ \Omega = R + \frac{1}{\dfrac{1}{(R + 5.0\ \Omega)} + \dfrac{1}{30\ \Omega}}$$

becomes

$$75\ \Omega = R + \frac{30\ \Omega(R + 5.0\ \Omega)}{(30\ \Omega + (R + 5.0\ \Omega))}$$

This becomes

$$R^2 - 10R - 2475 = 0.$$

When this equation is solved, we find the roots $55\ \Omega$ and $-45\ \Omega$. Because the resistance must be a positive quantity, we have $R = 55\ \Omega$.

Solutions

PROBLEM 87 Circuit Analysis and Kirchhoff's Laws

Before we begin...

1. The sum of the currents entering any junction must equal the sum of the currents leaving that same junction.

The sum of the potential differences across all of the elements that constitute any closed loop in a circuit is zero.

Solving the problem

2. Applying the junction rule, we have

$$I_2 = I_1 + I_3$$

3. Applying the loop rule to the left loop, we have

$$-(3\ \Omega)I_1 - (5\ \Omega)I_2 + (5.0\ \text{V}) = 0$$

For the right loop, we find

$$+(7\ \Omega)I_3 + (5\ \Omega)\ I_2 - (10.0\ \text{V}) = 0$$

4. Solving the equations by the method of substitution yields

$$I_1 = 0.141\ \text{A}$$

$$I_2 = 0.915\ \text{A}$$

$$I_3 = 0.774\ \text{A}$$

5. Evaluating the potential difference between a and b, we set the potential at b equal to zero and travel to a, summing the potential gains and drops as we progress.

$$V_b - V_a = 0 - 10.0\ \text{V} - (3.0\ \Omega)(0.141\ \text{A})$$
$$= -10.4\ \text{V}$$

If we are going with the current through a resistor, the potential drops. Going against the current through a resistor results in a potential gain.

PROBLEM 88 Capacitors

Before we begin...

1. The equivalent capacitance for capacitors in series is given by

$$\frac{1}{C_S} = \frac{1}{C_1} + \frac{1}{C_2} + \dots$$

2. The equivalent capacitance for capacitors in parallel is given by

$$C_P = C_1 + C_2 + \dots$$

3. Capacitors in series have the same charge on each capacitor.

Capacitors in parallel have the same potential difference across each.

4. The energy stored in a capacitor is computed by

$$U = \frac{1}{2}CV^2 \qquad \text{or} \qquad U = \frac{1}{2}\left(\frac{Q^2}{C}\right)$$

Solving the problem

5. The 3.0 μF capacitor and the 6.0 μF capacitor are connected in series.

$$\frac{1}{C_2} = \frac{1}{3.0\ \mu\text{F}} + \frac{1}{6.0\ \mu\text{F}}$$

$$C_2 = 2.0\ \mu\text{F}$$

6. The 2.0 μF capacitor and the 4.0 μF capacitor are connected in series

$$\frac{1}{C_3} = \frac{1}{2.0\ \mu\text{F}} + \frac{1}{4.0\ \mu\text{F}}$$

$$C_3 = 1.33\ \mu\text{F}$$

(continued on next page . . .)

7.

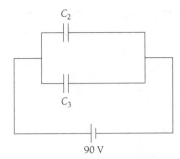

90 V

C_2 and C_3 are connected in parallel. The equivalent capacitance C_T is

$$2.0 \ \mu F + 1.3 \ \mu F = 3.3 \ \mu F$$

8. Because C_2 and C_3 are in parallel, they each have the same potential difference, 90 V. The charge on C_2 is

$$q = C_2 V = 180 \ \mu C$$

Because C_2 represents the series of capacitors, 3.0 μF and 6.0 μF, each of these have a charge of 180 μC. The charge on C_3 is

$$q = C_3 V = 120 \ \mu C$$

Because C_3 represents a series of capacitors, 2.0 μF and 4.0 μF, each of these have a charge of 120 μC. By using the relationship

$$V = Q/C$$

we compute the potential differences to be

voltage on 3.0 μF capacitor: $\dfrac{180 \ \mu C}{3.0 \ \mu F} = 60 \ V$

voltage on 6.0 μF capacitor: $\dfrac{180 \ \mu C}{6.0 \ \mu F} = 30 \ V$

voltage on 2.0 μF capacitor: $\dfrac{120 \ \mu C}{2.0 \ \mu F} = 60 \ V$

voltage on 4.0 μF capacitor: $\dfrac{120 \ \mu C}{4.0 \ \mu F} = 30 \ V$

9. The energy stored by this circuit of capacitors is

$$U = \frac{1}{2} C V^2 = \frac{1}{2} (3.33 \ \mu F)(90 \ V)^2$$

$$= 1.34 \times 10^{-2} \ J$$

PROBLEM 89 Inductors

Before we begin...

1. The potential difference created by an inductor is given by

$$V = L \frac{dI}{dt}$$

2. In the problem description the current has reached its steady-state value; therefore, the rate of change of the current is zero. No potential difference will be measured across the inductor.

3. The energy stored in the magnetic field of an inductor is computed by

$$U = \frac{1}{2} L I^2$$

where L is the inductance and I is the current passing through the inductor.

4. The given information is

$$V = 24.0 \ V \qquad R = 8.00 \ \Omega \qquad L = 4.00 \ H$$

Solving the problem

5. At equilibrium, the entire 24 volt drop occurs across the resistor. Using Ohm's law to compute the value of the equilibrium current, we have

$$I = V/R = (24.0 \ V)/(8.00 \ \Omega) = 3.00 \ A$$

6. The energy stored in the magnetic field of the inductor is

$$U = \frac{1}{2} L I^2 = \frac{1}{2} (4.00 \ H)(3.00 \ A)^2 = 18.0 \ J$$

PROBLEM 90 Circuits Containing Resistors, Inductors, and Capacitors

Before we begin...

1. The angular frequency ω_0 is related to the frequency f by

$$\omega_0 = 2\pi f$$

2. The given information is

$$f = 120 \text{ Hz} \qquad C = 8.00 \ \mu\text{F}$$

Solving the problem

3. The angular frequency ω_0 is

$$\omega_0 = 2\pi f = (6.28)(120 \text{ Hz}) = 754 \text{ s}^{-1}$$

4. Relating the angular frequency to the inductance and capacitance, we find

$$\omega_0 = \frac{1}{\sqrt{LC}}$$

The value of the inductance is thus

$$L = \frac{1}{(\omega_0{}^2 C)} = \ast\ast\ast \frac{1}{(754 \text{ s}^{-1})^2(8.00 \ \mu\text{F})}$$

$$= 0.220 \text{ H}$$

PROBLEM 91 Circuits Containing Resistors, Inductors, and Capacitors

Before we begin...

1. The relationship between ω_0, L, and C is

$$\omega_0 = \frac{1}{\sqrt{LC}}$$

2. The given information is

$$L = 2.18 \text{ H}$$

$$C = 6.00 \text{ nF}$$

Solving the problem

3. Setting the equations for ω_0 and for β equal to each other and solving for the limiting value of R, we find

$$\frac{R}{2L} = \frac{1}{\sqrt{LC}}$$

such that

$$R = (2)\sqrt{\frac{L}{C}} = (2)\sqrt{\frac{(2.18 \text{ H})}{6.00 \text{ nF}}} = 3.81 \times 10^4 \ \Omega$$

Before we begin...

1. The law of reflection states that the angle of incidence equals the angle of reflection as measured from the normal constructed to a reflecting surface.

2.

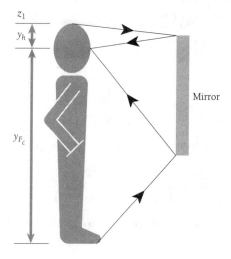

Mirror

Solving the problem

3. The mirror top must be placed $y_h/2$ or higher for the person still to see the top of his head because the angle of incidence must equal the angle of reflection. (We assume that the top of the head is in the same vertical plane with the eyes.)

4. The mirror bottom must be placed at least $y_f/2$ below the eyes for the person to see his feet.

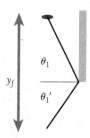

5. The minimum height of the mirror is calculated to be

$$h' = y_h/2 + y_f/2 = {}^1\!/_2(y_h + y_f) = {}^1\!/_2 h$$

Solutions

PROBLEM 93 Snell's Law

Before we begin...

1.

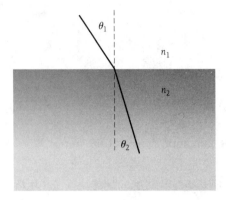

2. Snell's law states that the angle of refraction θ_2 depends on the properties of the two media and on the angle of incidence θ_1 from which the light passes.

$$n_1 \sin \theta_1 = n_2 \sin \theta_2$$

where n represents the index of refraction of the medium.

3. The given information is

$$\theta_1 = 30.0° \qquad n_1 = 1.00 \qquad n_2 = 1.33$$

Solving the problem

4. Applying Snell's law to solve for the angle at which the light will travel with respect to the normal while in the water, we use

$$n_1 \sin \theta_1 = n_2 \sin \theta_2$$

to evaluate

$$\sin \theta_2 = (n_1 \sin \theta_1)/n_2$$
$$= (1 \times 0.500)/(1.33) = 0.375$$

from which we compute

$$\theta_2 = \sin^{-1} (0.375) = 22.0°$$

PROBLEM 94 Snell's Law

Before we begin...

1. The given information is

$$\theta_1 = 45.0° \qquad n_1 = 1.00 \qquad n_2 = 1.52$$

Solving the problem

2. Using Snell's law and the given information from above to calculate θ_2, we find

$$n_1 \sin \theta_1 = n_2 \sin \theta_2$$

$$\sin \theta_2 = (n_1 \sin \theta_1)/n_2 = 0.465$$

so

$$\theta_2 = \sin^{-1} (0.465) = 27.7°$$

3. Examining triangle ABC, we observe that

$$\tan \theta_2 = (10 \text{ cm} - x)/(10 \text{ cm})$$

4. Using this relationship to solve for x gives us

$$x = 10 \text{ cm} - (10 \text{ cm}) \tan 27.7° = 4.74 \text{ cm}$$

Saunders Core Concepts in Physics Workbook

PROBLEM 95 Total Internal Reflection

Before we begin...

1. The critical angle is the angle of incidence θ_C such that the angle of refraction θ_2 is 90°. The refracted light will move parallel to the boundary. This is possible only when light attempts to move from a medium with a higher index of refraction to a medium with a lower index of refraction.

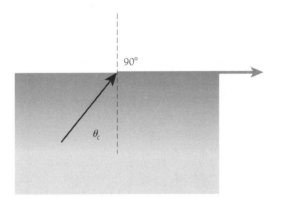

2. The given information is

$$n_1 = 1.50 \qquad n_2 = 1.33 \qquad \theta_2 = 90°$$

Solving the problem

3. Applying Snell's law and the definition of the critical angle to compute θ_C, we find

$$n_1 \sin \theta_C = n_2 \sin 90°$$

Solving for θ_C, we have

$$\sin \theta_C = n_2 (1.00)/n_1 = (1.33)/(1.50) = 0.887$$

from which we compute

$$\theta_C = \sin^{-1} (0.887) = 62.5°$$

PROBLEM 96 Flat and Spherical Mirrors

Before we begin...

1. The lateral magnification is defined as the ratio of the height of the image to the height of the object. For spherical mirrors, it can be computed as

$$M = -d_i/d_o$$

2. The mirror equation is

$$\frac{1}{f} = \frac{1}{d_o} + \frac{1}{d_i}$$

where f is the focal length, d_o the object distance, and d_i the image distance from the mirror.

3. Upright images are virtual. They appear behind the viewing surface of the mirror. The image distance of a virtual image is negative. A concave (or converging) mirror has a positive focal length.

4. The given information is

$$M = +4 \qquad\qquad f = +40 \text{ cm}$$

Solving the problem

5. Substituting the relation

$$M = -\frac{d_i}{d_o}$$

into the mirror equation yields

$$\frac{1}{f} = \frac{1}{d_o} + \frac{1}{(-Md_o)} \qquad \text{or} \qquad d_o = f\left(1 - \frac{1}{M}\right)$$

Substituting the values $f = +40$ cm and $M = +4$, we find

$$d_o = 30 \text{ cm}$$

PROBLEM 97 Flat and Spherical Mirrors

Before we begin...

1. The focal length of a spherical mirror is equal in magnitude to half the mirror's radius of curvature

$$f = \pm R/2$$

2. The focal length of a convex mirror is negative. The center of curvature is on the virtual side of the mirror.

3. The given information is

$$d_o = +15 \text{ cm} \qquad R = -3.0 \text{ cm}$$

Solving the problem

4. The focal length of the mirror is

$$f = -R/2 = -(3.0 \text{ cm})/2 = -1.5 \text{ cm}$$

5. The image distance d_i is

$$\frac{1}{f} = \frac{1}{d_o} + \frac{1}{d_i}$$

or

$$\frac{1}{-1.5 \text{ cm}} = \frac{1}{15 \text{ cm}} + \frac{1}{d_i}$$

Evaluating the expression, we find

$$d_i = (-15 \text{ cm})/11 = -1.36 \text{ cm}$$

6. The equation for lateral magnification allows us to find

$$M = -d_i/d_o = -(-1.36 \text{ cm})/(+15 \text{ cm}) = +0.091$$

That is, the image is upright, about 9% of the size of the object, and located 1.36 cm behind the surface of the ornament.

PROBLEM 98 Thin Lenses

Before we begin...

1. If an image is projected onto a screen, it must be a real image. Virtual images cannot be projected.

2. The thin lens equation states

$$\frac{1}{f} = \frac{1}{d_o} + \frac{1}{d_i}$$

where f is the focal length, d_o the object distance, and d_i the image distance from the lens.

3. Lateral magnification is defined as the ratio of the height of the image to the height of the object. For thin lenses, it can be computed as

$$M = -d_i/d_o$$

4. The given information is

$$d_o = +32 \text{ cm} \qquad d_i = +8.0 \text{ cm}$$

Solving the problem

5. Substituting into the thin lens equation, we find

$$\frac{1}{f} = \frac{1}{32 \text{ cm}} + \frac{1}{8.0 \text{ cm}}$$

or

$$f = (+32 \text{ cm})/5 = +6.4 \text{ cm}$$

Because the focal length is positive, the lens is converging.

6. The lateral magnification is calculated as

$$M = -(d_i)/(d_f) = -(8.0 \text{ cm})/(32 \text{ cm}) = -0.25$$

In other words, the image is inverted and one quarter the size of the object.

Saunders Core Concepts in Physics Workbook

Before we begin...

1. The magnification of any image is defined as the ratio of the image height h_i to the object height h_o

$$M = -\frac{h_i}{h_o}$$

For a thin lens, the magnification is always equal to $-d_i/d_o$, where d_i is the image distance and d_o is the object distance.

2. The thin lens equation relates d_i and d_o to the focal length f of the lens:

$$\frac{1}{f} = \frac{1}{d_i} + \frac{1}{d_o}$$

Solving the problem

3. The object's height is $h_o = 2.40$ cm $= 0.024$ m, and the image height is $h_i = -1.80$ m (negative because the real image of a lens is always inverted relative to the real object that cast it). Thus the magnification is

$$M = h_i/h_o = -75$$

Knowing the magnification, we can now predict that

$$d_i/d_o = -M = 75$$

4. Together, the object distance d_o and the image distance d_i make up the entire distance from the slide to the screen, which the problem tells us is three meters

$$d_o + d_i = 3 \text{ m}$$

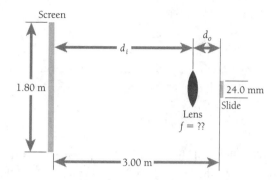

5. Solving the above two equations together, we find that

$$d_o = 0.0395 \text{ m} = 3.95 \text{ cm}$$

$$d_i = 2.96 \text{ m}$$

6. Substituting these values of d_o and d_i into the thin lens equation,

$$\frac{1}{f} = \frac{1}{d_i} + \frac{1}{d_o}$$

we can determine that the focal length of the lens is

$$f = 0.0390 \text{ m} = 3.90 \text{ m}$$

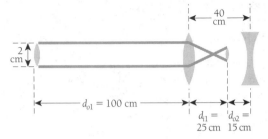

Before we begin...

1. For each lens, the magnification M is given by the ratio of the image and object distances

$$M = -\frac{d_i}{d_o}$$

2. The object of the first lens is the coin. It is a real object, at a distance $d_{o1} = 100$ cm from the lens.

3. The converging lens has focal length $f_1 = 20$ cm. The diverging lens has focal length $f_2 = -30$ cm.

Solving the problem

4. The object and image distances of the first lens are related by the thin lens equation

$$\frac{1}{d_{o1}} + \frac{1}{d_{i1}} = \frac{1}{f_1}$$

Using the known values $d_{o1} = 100$ cm and $f_1 = 20$ cm, we can solve for the image distance

$$\frac{1}{d_{i1}} = 0.04 \text{ cm}^{-1} \quad \text{or} \quad d_{i1} = 25 \text{ cm}$$

Thus the first lens casts a real image of the coin, 25 cm to the right of the lens.

5. The magnification of the first image is

$$M_1 = -\frac{d_{i1}}{d_{o1}} = -0.25$$

In other words, the image is inverted (since M is negative) and one-quarter the size of the original coin. It has a diameter of 0.5 cm.

6. The image of the first lens is the object of the second lens. It is a real object (since light actually passes through this point before encountering the second lens) and is 15 cm to the left of the second lens, as shown in the diagram.

Thus the diverging lens has object distance $d_{o2} = 15$ cm (positive because this is a real object).

7. Knowing this and the focal length $f_2 = -30$ cm, we can again use the thin lens equation

$$\frac{1}{d_{o2}} + \frac{1}{d_{i2}} = \frac{1}{f_2}$$

to find the position of the final image

$$\frac{1}{d_{i2}} = -0.10 \text{ cm}^{-1} \quad \text{or} \quad d_{i2} = -10 \text{ cm}$$

The final image of the coin is thus a virtual image (because d_{i2} is negative), 10 cm to the left of the diverging lens.

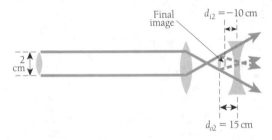

8. The second lens has a magnification of

$$M_2 = -\frac{d_{i2}}{d_{o2}} = 0.667$$

Recall that the object of the second lens (the image of the first lens) was upside-down relative to the coin, and had diameter 0.5 cm. The image of the second lens is two-thirds that size, with diameter 0.33 cm. Finally, since M_2 is positive, its image is upright relative to its object—and thus inverted relative to the original coin.

Reference Tables

Some Fundamental Constants[a]

Quantity	Symbol	Value[b]
Atomic mass unit	u	$1.660\ 540\ 2(10) \times 10^{-27}$ kg $931.434\ 32(28)$ MeV/c^2
Avogadro's number	N_A	$6.002\ 136\ 7(36) \times 10^{23}$ (mol)$^{-1}$
Bohr magneton	$\mu_B = \dfrac{e\hbar}{2m_e}$	$9.274\ 015\ 4(31) \times 10^{-24}$ J/T
Bohr radius	$a_0 = \dfrac{\hbar^2}{m_e e^2 k_e}$	$0.529\ 177\ 249(24) \times 10^{-10}$ m
Boltzmann's constant	$k_B = R/N_A$	$1.380\ 658(12) \times 10^{-23}$ J/K
Compton wavelength	$\lambda_c = \dfrac{h}{m_e c}$	$2.426\ 310\ 58(22) \times 10^{-12}$ m
Deuteron mass	m_d	$3.343\ 586\ 0(20) \times 10^{-27}$ kg $2.013\ 553\ 214(24)$ u
Electron mass	m_e	$9.109\ 389\ 7(54) \times 10^{-31}$ kg $5.485\ 799\ 03(13) \times 10^{-4}$ u $0.510\ 999\ 06(15)$ MeV/c^2
Electron-volt	eV	$1.602\ 177\ 33(49) \times 10^{-19}$ J
Electron charge	e	$1.602\ 177\ 33(49) \times 10^{-19}$ C
Gas constant	R	$8.314\ 510(70)$ J/K $\cdot$ mol
Gravitational constant	G	$6.672\ 59(85) \times 10^{-11}$ N$\cdot$m^2/kg^2
Hydrogen ground state	$E_0 = \dfrac{m_e e^4 k_e^2}{2\hbar^2} = \dfrac{e^2 k_e}{2a_0}$	$13.605\ 698(40)$ eV
Josephson frequency-voltage ratio	$2e/h$	$4.835\ 976\ 7(14) \times 10^{14}$ Hz/V

Some Fundamental Constants[a]   (continued)

Quantity	Symbol	Value[b]
Magnetic flux quantum	$\Phi_0 = \dfrac{h}{2e}$	$2.067\ 834\ 61(61) \times 10^{-15}$ Wb
Neutron mass	m_n	$1.674\ 928\ 6(10) \times 10^{-27}$ kg $1.008\ 664\ 904(14)$ u $939.565\ 63(28)$ MeV/c^2
Nuclear magneton	$\mu_n = \dfrac{e\hbar}{2m_P}$	$5.050\ 786\ 6(17) \times 10^{-27}$ J/T
Permeability of free space	μ_0	$4\pi \times 10^{-7}$ N/A^2 (exact)
Permittivity of free space	$\epsilon_0 = 1/\mu_0 c^2$	$8.854\ 187\ 817 \times 10^{-12}$ C^2/N·m^2 (exact)
Planck's constant	h $\hbar = h/2\pi$	$6.626\ 075(40) \times 10^{-34}$ J·s $1.054\ 572\ 66(63) \times 10^{-34}$ J·s
Proton mass	m_p	$1.672\ 623(10) \times 10^{-27}$ kg $1.007\ 276\ 470(12)$ u $938.272\ 3(28)$ MeV/c^2
Quantized Hall resistance	h/e^2	$25812.805\ 6(12)$ Ω
Rydberg constant	R_H	$1.097\ 373\ 153\ 4(13) \times 10^7$ m^{-1}
Speed of light in vacuum	c	$2.997\ 924\ 58 \times 10^8$ m/s (exact)

[a]These constants are the values recommended in 1986 by CODATA, based on a least-squares adjustment of data from different measurements. For a more complete list, see Cohen, E. Richard, and Barry N. Taylor, *Rev. Mod. Phys.* 59:1121, 1987

[b]The numbers in parentheses for the values below represent the uncertainties in the last decimal places.

Saunders Core Concepts in Physics Workbook

Solar System Data

Body	Mass (kg)	Mean Radius (m)	Period (s)	Distance from Sun (m)
Mercury	3.18×10^{23}	2.43×10^{6}	7.60×10^{6}	5.79×10^{10}
Venus	4.88×10^{24}	6.06×10^{6}	1.94×10^{7}	1.08×10^{11}
Earth	5.98×10^{24}	6.37×10^{6}	3.156×10^{7}	1.496×10^{11}
Mars	6.42×10^{23}	3.37×10^{6}	5.94×10^{7}	2.28×10^{11}
Jupiter	1.90×10^{27}	6.99×10^{7}	3.74×10^{8}	7.78×10^{11}
Saturn	5.68×10^{26}	5.85×10^{7}	9.35×10^{8}	1.43×10^{12}
Uranus	8.68×10^{25}	2.33×10^{7}	2.64×10^{9}	2.87×10^{12}
Neptune	1.03×10^{26}	2.21×10^{7}	5.22×10^{9}	4.50×10^{12}
Pluto	$\approx 1.4 \times 10^{22}$	$\approx 1.5 \times 10^{6}$	7.82×10^{9}	5.91×10^{12}
Moon	7.36×10^{22}	1.74×10^{6}	—	—
Sun	1.991×10^{30}	6.96×10^{8}	—	—

Physical Data Often Used[a]

Average Earth-Moon distance	3.84×10^{8} m
Average Earth-Sun distance	1.496×10^{11} m
Average radius of the Earth	6.37×10^{6} m
Density of air (20°C and 1 atm)	1.29 kg/m^3
Density of water (20°C and 1 atm)	1.00×10^{3} kg/m^3
Free-fall acceleration	9.80 m/s^2
Mass of the Earth	5.98×10^{24} kg
Mass of the Moon	7.36×10^{22} kg
Mass of the Sun	1.99×10^{30} kg
Standard atmospheric pressure	1.013×10^{5} Pa

[a]These are the values of the constants as used in the text.

Appendix A

Some Prefixes for Powers of Ten

Power	Prefix	Abbreviation
10^{-18}	atto	a
10^{-15}	femto	f
10^{-12}	pico	p
10^{-9}	nano	n
10^{-6}	micro	μ
10^{-3}	milli	m
10^{-2}	centi	c
10^{-1}	deci	d
10^{1}	deka	da
10^{2}	hecto	h
10^{3}	kilo	k
10^{6}	mega	M
10^{9}	giga	G
10^{12}	tera	T
10^{15}	peta	P
10^{18}	exa	E

Standard Abbreviations and Symbols of Units

Abbreviation	Unit
A	ampere
Å	angstrom
u	atomic mass unit
atm	atmosphere
Btu	British thermal unit
C	coulomb
°C	degree Celsius
cal	calorie
deg	degree (angle)
eV	electron volt
°F	degree Fahrenheit
F	farad
ft	foot
G	gauss
g	gram
H	henry
h	hour
hp	horsepower
Hz	hertz
in.	inch
J	joule
K	kelvin
kcal	kilocalorie
kg	kilogram
kmol	kilomole
lb	pound
m	meter
min	minute
N	newton
Pa	pascal
rev	revolution

(*continued on next page . . .*)

Saunders Core Concepts in Physics Workbook

Standard Abbreviations and Symbols of Units
(continued)

Abbreviation	Unit
s	second
T	tesla
V	volt
W	watt
Wb	weber
μm	micrometer
Ω	ohm

Mathematical Symbols Used in the Text and Their Meaning

Symbol	Meaning
$=$	is equal to
$\equiv$	is defined as
$\neq$	is not equal to
$\propto$	is proportional to
$>$	is greater than
$<$	is less than
$\gg (\ll)$	is much greater (less) than
$\approx$	is approximately equal to
Δx	the change in x
$\sum_{i=1}^{N} x_i$	the sum of all quantities x_i from $i = 1$ to $i = N$
$\lvert x \rvert$	the magnitude of x (always a positive quantity)
$\Delta x \rightarrow 0$	Δx approaches zero
$\dfrac{dx}{dt}$	the derivative of x with respect to t
$\dfrac{\partial x}{\partial t}$	the partial derivative of x with respect to t
$\displaystyle\int$	integral

Useful Conversions

Length

12 in. = 1 ft

3 ft = 1 yd

1 yd = 0.9144 m

$1 \text{ Å} = 10^{-10}$ m

$1 \ \mu\text{m} = 1 \ \mu = 10^{-6}$ m $= 10^4$ Å

1 lightyear $= 9.461 \times 10^{15}$ m

Area

$1 \text{ m}^2 = 10^4 \text{ cm}^2 = 10.76 \text{ ft}^2$

$1 \text{ ft}^2 = 0.0929 \text{ m}^2 = 144 \text{ in.}^2$

$1 \text{ in.}^2 = 6.452 \text{ cm}^2$

Volume

$1 \text{ m}^3 = 10^6 \text{ cm}^3 = 6.102 \times 10^4 \text{ in.}^3$

$1 \text{ ft}^3 = 1728 \text{ in.}^3 = 2.83 \times 10^{-2} \text{ m}^3$

$1 \text{ liter} = 1000 \text{ cm}^3 = 1.0576 \text{ qt} = 0.0353 \text{ ft}^3$

$1 \text{ ft}^3 = 7.481 \text{ gal} = 28.32 \text{ liters} = 2.832 \times 10^{-2} \text{ m}^3$

$1 \text{ gal} = 3.786 \text{ liters} = 231 \text{ in.}^3$

Mass

1000 kg = 1 t (metric ton)

Velocity

1 mi/min = 60 mi/h = 88 ft/s

Acceleration

$1 \text{ m/s}^2 = 3.28 \text{ ft/s}^2 = 100 \text{ cm/s}^2$

$1 \text{ ft/s}^2 = 0.3048 \text{ m/s}^2 = 30.48 \text{ cm/s}^2$

Pressure

$1 \text{ bar} = 10^5 \text{ N/m}^2 = 14.50 \text{ lb/in.}^2$

Energy

931.5 MeV/c^2 is equivalent to 1 u

Power

1 hp = 550 ft $\cdot$ lb/s = 0.746 kW

1 W = 1 J/s = 0.738 ft $\cdot$ lb/s

1 Btu/h = 0.293 W

The Greek Alphabet

Alpha	A	α
Beta	B	β
Gamma	Γ	γ
Delta	Δ	δ
Epsilon	E	ϵ
Zeta	Z	ζ
Eta	H	η
Theta	Θ	θ
Iota	I	ι
Kappa	K	κ
Lambda	Λ	λ
Mu	M	μ
Nu	N	ν
Xi	Ξ	ξ
Omicron	O	o
Pi	Π	π
Rho	P	ρ
Sigma	Σ	σ
Tau	T	τ
Upsilon	Υ	υ
Phi	Φ	ϕ
Chi	X	χ
Psi	Ψ	ψ
Omega	Ω	ω

Conversion Factors

LENGTH	m	cm	km	in.	ft	mi
1 meter	1	10^2	10^{-3}	39.37	3.281	6.214×10^{-4}
1 centimeter	10^{-2}	1	10^{-5}	0.3937	3.281×10^{-2}	6.214×10^{-6}
1 kilometer	10^3	10^5	1	3.937×10^4	3.281×10^3	0.6214
1 inch	2.540×10^{-2}	2.540	2.540×10^{-5}	1	8.333×10^{-2}	1.578×10^{-5}
1 foot	0.3048	30.48	3.048×10^{-4}	12	1	1.894×10^{-4}
1 mile	1609	1.609×10^5	1.609	6.336×10^4	5280	1

MASS	kg	g	slug	u
1 kilogram	1	10^3	6.852×10^{-2}	6.024×10^{26}
1 gram	10^{-3}	1	6.852×10^{-5}	6.024×10^{23}
1 slug	14.59	1.459×10^4	1	8.789×10^{27}
1 atomic mass unit	1.660×10^{-27}	1.660×10^{-24}	1.137×10^{-28}	1

TIME	s	min	h	day	year
1 second	1	1.667×10^{-2}	2.778×10^{-4}	1.157×10^{-5}	3.169×10^{-8}
1 minute	60	1	1.667×10^{-2}	6.994×10^{-4}	1.901×10^{-6}
1 hour	3600	60	1	4.167×10^{-2}	1.141×10^{-4}
1 day	8.640×10^4	1440	24	1	2.738×10^{-3}
1 year	3.156×10^7	5.259×10^5	8.766×10^3	365.2	1

SPEED	m/s	cm/s	ft/s	mi/h
1 meter/second	1	10^2	3.281	2.237
1 centimeter/second	10^{-2}	1	3.281×10^{-2}	2.237×10^{-2}
1 foot/second	0.3048	30.48	1	0.6818
1 mile/hour	0.4470	44.70	1.467	1

Note: 1 mi/min = 60 mi/h = 88 ft/s.

FORCE	N	dyn	lb
1 newton	1	10^5	0.2248
1 dyne	10^{-5}	1	2.248×10^{-6}
1 pound	4.448	4.448×10^5	1

WORK, ENERGY, HEAT	J	erg	ft · lb
1 joule	1	10^7	0.7376
1 erg	10^{-7}	1	7.376×10^{-8}
1 ft · lb	1.356	1.356×10^7	1
1 eV	1.602×10^{-19}	1.602×10^{-12}	1.182×10^{-19}
1 cal	4.186	4.186×10^7	3.087
1 Btu	1.055×10^3	1.055×10^{10}	7.779×10^2
1 kWh	3.600×10^6	3.600×10^{13}	2.655×10^6

	eV	cal	Btu	kWh
1 joule	6.242×10^{18}	0.2389	9.481×10^{-4}	2.778×10^{-7}
1 erg	6.242×10^{11}	2.389×10^{-8}	9.481×10^{-11}	2.778×10^{-14}
1 ft · lb	8.464×10^{18}	0.3239	1.285×10^{-3}	3.766×10^{-7}
1 eV	1	3.827×10^{-20}	1.519×10^{-22}	4.450×10^{-26}
1 cal	2.613×10^{19}	1	3.968×10^{-3}	1.163×10^{-6}
1 Btu	6.585×10^{21}	2.520×10^2	1	2.930×10^{-4}
1 kWh	2.247×10^{25}	8.601×10^5	3.413×10^2	1

PRESSURE			
	Pa	dyn/cm^2	atm
1 pascal	1	10	9.869×10^{-6}
1 dyne/centimeter^2	10^{-1}	1	9.869×10^{-7}
1 atmosphere	1.013×10^5	1.013×10^6	1
1 centimeter mercury*	1.333×10^3	1.333×10^4	1.316×10^{-2}
1 pound/inch^2	6.895×10^3	6.895×10^4	6.805×10^{-2}
1 pound/foot^2	47.88	4.788×10^2	4.725×10^{-4}
	cm Hg	lb/in.^{2}	lb/ft^2
1 newton/meter^2	7.501×10^{-4}	1.450×10^{-4}	2.089×10^{-2}
1 dyne/centimeter^2	7.501×10^{-5}	1.450×10^{-5}	2.089×10^{-3}
1 atmosphere	76	14.70	2.116×10^3
1 centimeter mercury*	1	0.1943	27.85
1 pound/inch^2	5.171	1	144
1 pound/foot^2	3.591×10^{-2}	6.944×10^{-3}	1

*At 0°C and at a location where the acceleration due to gravity has its "standard" value, 9.80665 m/s^2.

Symbols, Dimensions, and Units of Physical Quantities

Quantity	Common Symbol	Unit*	Dimensions†	Unit in Terms of Base SI Units
Acceleration	a	m/s^2	L/T^2	m/s^2
Amount of substance	n	mole		mol
Angle	θ, ϕ	radian (rad)	1	
Angular acceleration	$\vec{\alpha}$	rad/s^2	T^{-2}	s^{-2}
Angular frequency	ω	rad/s	T^{-1}	s^{-1}
Angular momentum	L	kg·m^2/s	ML^2/T	kg·m^2/s
Angular velocity	$\vec{\omega}$	rad/s	T^{-1}	s^{-1}
Area	A	m^2	L^2	m^2
Atomic number	Z			
Capacitance	C	farad (F)(= Q/V)	Q^{2}T^2/ML^2	A^2·s^4/kg·m^2
Charge	q, Q, e	coulomb (C)	Q	A·s

Symbols, Dimensions, and Units of Physical Quantities   (continued)

Quantity	Common Symbol	Unit*	Dimensions†	Unit in Terms of Base SI Units
Charge density				
Line	λ	C/m	Q/L	A·s/m
Surface	σ	C/m^2	Q/L^2	$A·s/m^2$
Volume	ρ	C/m^3	Q/L^3	$A·s/m^3$
Conductivity	σ	$1/\Omega·m$	Q^2T/ML^3	$A^2·s^3/kg·m^3$
Current	I	AMPERE	Q/T	A
Current density	$\vec{J}$	A/m^2	Q/T^2	A/m^2
Density	ρ	kg/m^3	M/L^3	kg/m^3
Dielectric constant	κ			
Displacement	s	METER	L	m
Distance	d, h			
Length	ℓ, L			
Electric dipole moment	$\vec{p}$	C·m	QL	A·s·m
Electric field	$\vec{E}$	V/m	ML/QT^2	$kg·m/A·s^3$
Electric flux	Φ	V·m	ML^3/QT^2	$kg·m^3/A·s^3$
Electromotive force	$\vec{\varepsilon}$	volt (V)	ML^2/QT^2	$kg·m^2/A·s^3$
Energy	E, U, K	joule (J)	ML^2/T^2	$kg·m^2/s^2$
Entropy	S	J/K	$ML^2/T^2·K$	$kg·m^2/s^2·K$
Force	$\vec{F}$	newton (N)	ML/T^2	$kg·m/s^2$
Frequency	f, v	hertz (Hz)	T^{-1}	s^{-1}
Heat	Q	joule (J)	ML^2/T^2	$kg·m^2/s^2$
Inductance	L	henry (H)	ML^2/Q^2	$kg·m^2/A^2·s^2$
Magnetic dipole moment	$\vec{\mu}$	N·m/T	QL^2/T	$A·m^2$
Magnetic field	$\vec{B}$	tesla (T) (=Wb/m^2)	M/QT	$kg/A·s^2$
Magnetic flux	Φ_m	weber (Wb)	ML^2/QT	$kg·m^2/A·s^2$
Mass	m, M	KILOGRAM	M	kg
Molar specific heat	C	J/mol·K		$kg·m^2/s^2·mol·K$
Moment of inertia	I	$kg·m^2$	ML^2	$kg·m^2$
Momentum	$\vec{p}$	kg·m/s	ML/T	kg·m/s
Period	T	s	T	s
Permeability of space	μ_0	N/A^2 (=H/m)	ML/Q^2T	$kg·m/A^2·s^2$
Permittivity of space	ϵ_0	$C^2/N·m^2$ (=F/m)	Q^2T^2/ML^3	$A^2·s^4/kg·m^3$

Symbols, Dimensions, and Units of Physical Quantities   (continued)

Potential (voltage)	V	volt (V)(=J/C)	ML^2/QT^2	$kg \cdot m^2/A \cdot s^3$
Power	P	watt (W)(=J/s)	ML^2/T^3	$kg \cdot m^2/s^3$
Pressure	P, p	pascal (Pa) = (N/m^2)	M/LT^2	$kg/m \cdot s^2$
Resistance	R	ohm (Ω)(=V/A)	ML^2/Q^2T	$kg \cdot m^2/A^2 \cdot s^3$
Specific heat	c	J/kg·K	$L^2/T^2 \cdot K$	$m^2/s^2 \cdot K$
Temperature	T	KELVIN	K	K
Time	t	SECOND	T	s
Torque	$\vec{\tau}$	N·m	ML^2/T^2	$kg \cdot m^2/s^2$
Speed	v	m/s	L/T	m/s
Volume	V	m^3	L^3	m^3
Wavelength	λ	m	L	m
Work	W	joule (J)(=N·m)	ML^2/T^2	$kg \cdot m^2/s^2$

*The base SI units are given in uppercase letters.

†The symbols M, L, T, and Q denote mass, length, time, and charge, respectively.

SI Base Units

Base Quantity	SI BASE UNIT	
	Name	Symbol
Length	meter	m
Mass	kilogram	kg
Time	second	s
Electric current	ampere	A
Temperature	kelvin	K
Amount of substance	mole	mol
Luminous intensity	candela	cd

Some Derived SI Units

Quantity	Name	Symbol	Expression in Terms of Base Units	Expression in Terms of Other SI Units
Plane angle	radian	rad	m/m	
Frequency	hertz	Hz	s^{-1}	
Force	newton	N	$kg \cdot m/s^2$	J/m
Pressure	pascal	Pa	$kg/m \cdot s^2$	N/m^2
Energy: work	joule	J	$kg \cdot m^2/s^2$	$N \cdot m$
Power	watt	W	$kg \cdot m^2/s^3$	J/s
Electric charge	coulomb	C	$A \cdot s$	
Electric potential (emf)	volt	V	$kg \cdot m^2/A \cdot s^3$	W/A
Capacitance	farad	F	$A^2 \cdot s^4/kg \cdot m^2$	C/V
Electric resistance	ohm	Ω	$kg \cdot m^2/A^2 \cdot s^3$	V/A
Magnetic flux	weber	Wb	$kg \cdot m^2/A \cdot s^2$	$V \cdot s$
Magnetic field intensity	tesla	T	$kg/A \cdot s^2$	Wb/m^2
Inductance	henry	H	$kg \cdot m^2/A^2 \cdot s^2$	Wb/A

Saunders Core Concepts in Physics Workbook

Correlation Guides

Problem Correlation Guide for Saunders Core Concepts in Physics CD-ROM

	Screen Number	Pop Questions	Step Problems	Workbook Problems
Module 2: Vectors	2.2	×		×
	2.3			
	2.4	×	×	×
	2.5	×		××
	2.6	×	×	×
	2.7	×	×	××
	2.8			
Module 3: Kinematics	3.2	×		×
	3.3	×	×	×
	3.4	×	×	×
	3.5	×		××
	3.6	×	×	××
	3.7	×		×
	3.8	×		
Module 4: Forces	4.2	×		×
	4.3	×		×
	4.4	×	×	×
	4.5	×	×	×
	4.6		×	××
	4.7	×		×
	4.8			×
	4.9	×		

(continued on next page . . .)

	Screen Number	Pop Questions	Step Problems	Workbook Problems
Module 5: **Work and Energy**	5.2	×	×	×
	5.3	×	×	××
	5.4	×		×
	5.5			
	5.6	×		×
	5.7	×		×
	5.8	×	×	×
	5.9	×		×
	5.10	×		
Module 6: **Linear Momentum**	6.2	×	×	
	6.3	×	×	×
	6.4	×		×
	6.5	×		××
	6.6	×		×
	6.7	×		×
	6.8	×	×	××
	6.9	×		
Module 7: **Rotational Mechanics**	7.2	×	×	×
	7.3	×		×
	7.4	×		×
	7.5	×	×	×
	7.6	×		×
	7.7	×	×	×
	7.8	×		×
	7.9	×		×
	7.10			

(continued on next page . . .)

Saunders Core Concepts in Physics Workbook

	Screen Number	Pop Questions	Step Problems	Workbook Problems
Module 8: Simple Harmonic Motion and Waves	8.2	✕		
	8.3	✕		✕
	8.4			
	8.5	✕		
	8.6	✕		✕
	8.7			
	8.8	✕	✕	✕✕
	8.9			✕
	8.10		✕	✕
	8.11	✕		✕
	8.12		✕	
	8.13	✕		
	8.14	✕		
Module 9: Wave Behavior	9.2	✕	✕	✕
	9.3	✕		
	9.4			
	9.5			✕
	9.6	✕	✕	
	9.7			✕
	9.8	✕		✕
	9.9		✕	✕✕
	9.10	✕		
	9.11	✕		✕
	9.12	✕		

(continued on next page . . .)

Appendix B

	Screen Number	Pop Questions	Step Problems	Workbook Problems
Module 10: Thermodynamics	10.2	×		
	10.3	×		×
	10.4	×		
	10.5	×	×	×
	10.6	×	×	×
	10.7	×		
	10.8			
	10.9	×	×	××
	10.10	×		
	10.11	×		××
	10.12			
Module 11: The Electric Field	11.2	×	×	
	11.3	×		
	11.4	×		×
	11.5	×	×	×
	11.6	×	×	××
	11.7	×		×
	11.8	×		×
	11.9	×		××
	11.10	×		
Module 12: The Magnetic Field	12.2		×	
	12.3	×		××
	12.4	×		×
	12.5	×		×
	12.6	×		
	12.7	×	×	××
	12.8	×		×
	12.9	×	×	
	12.10	×		
	12.11	×		

(continued on next page . . .)

	Screen Number	Pop Questions	Step Problems	Workbook Problems
Module 13: **Electric Circuits**	13.2	✕	✕	
	13.3	✕		✕
	13.4	✕		✕✕
	13.5	✕	✕	✕
	13.6	✕		✕
	13.7	✕	✕	✕✕
	13.8	✕		
Module 14: **Geometric Optics**	14.2	✕		
	14.3	✕		✕
	14.4	✕	✕	✕✕
	14.5	✕		✕
	14.6	✕		
	14.7	✕	✕	✕✕
	14.8	✕	✕	✕✕✕
	14.9			

Physics Textbook Correlation Guide

This guide provides information about sections in Physics textbooks that correspond to screens in the *Saunders Core Concepts in Physics CD-ROM*.

Textbooks include:
Serway: *Physics for Scientists and Engineers, 4/e (PSE)*
Serway: *Principles of Physics, 2/e (POP)*
Halliday, Resnick, and Walker: *Fundamentals of Physics, 5/e (FOP)*
Young and Freedman: *University Physics, 9/e (UP)*
Tipler: *Physics for Scientists and Engineers 3/e (PSE)*

Screen Number (CD-ROM)	Serway: PSE, 4/e	Serway: POP, 2/e	Halliday, Resnick, and Walker: FOP, 5/e	Young and Freedman: UP, 9/e	Tipler: PSE, 3/e
1.1	n/a	n/a	n/a	n/a	n/a
1.2	n/a	n/a	Problem Solving Tactic 2-1	n/a	n/a
1.3	Preface	1.8	5-2	n/a	n/a
1.4	n/a	1.8+	Problem Solving Tactic 2-2	n/a	n/a
1.5	1.4, 1.6	1.3, 1.5	n/a	1-5, 1-7	1-5
1.6	n/a	1.8+	n/a	n/a	n/a
1.7	n/a	1.8, 1.3, 1.5 +	Problem Solving Tactic 2-2	n/a	n/a
1.8	n/a	n/a	n/a	n/a	n/a
2.1	n/a	n/a	n/a	n/a	n/a
2.2	3.1	1.7	n/a	n/a	n/a
2.3	3.2	1.9	3-1	1-8	3-1
2.4	3.3	1.10	3-2	1-8	3-2
2.5	3.4	1.11	3-3, 3-4	1-9, 1-10	3-3
2.6	7.2	1.10	3-7	1-11	6-3
2.7	11.2	1.10	3-7	1-11	8-7
2.8	n/a	n/a	n/a	n/a	n/a
3.1	n/a	n/a	n/a	n/a	n/a
3.2	2.1	2.1, 2.2	2-2, 2-3	2-2, 2-3	2-1

(continued on next page . . .)

Screen Number (CD-ROM)	Serway: PSE, 4/e	Serway: POP, 2/e	Halliday, Resnick, and Walker: FOP, 5/e	Young and Freedman: UP, 9/e	Tipler: PSE, 3/e
3.3	2.2	2.3	2-4, 2-5	2-3, 2-4	2-2, 2-3
3.4	2.4	2.5	2-6	2-5	2-4
3.5	4.1, 4.2, 4.3	3.3	4-5, 4-6	3-4	3-7
3.6	4.4, 4.5	3.4, 5.3	4-7	3-5	3-8
3.7	4.6	1.7	4-8	3-6	3-6
3.8	n/a	n/a	n/a	n/a	n/a
4.1	n/a	n/a	n/a	n/a	n/a
4.2	5.1, 5.2, 6.6	4.2	5-2, 5-3	4-3	4-1, 4-5
4.3	5.3, 5.5	4.3	5-6	4-4, 4-5	4-2, 4-3
4.4	5.4	4.4	5-5	4-4	4-2
4.5	5.6	4.6	5-7	4-6	4-4
4.6	5.7	4.7	5-5	4-7, 4-8	4-6
4.7	6.1	5.2	4-7	5-5	4-6 +
4.8	6.3	n/a	n/a	4-3	5-4
4.9	n/a	4.2	n/a	n/a	n/a
5.1	n/a	n/a	n/a	n/a	n/a
5.2	7.1	6.1, 6.3	7-2	6-2	6-1, 6-2
5.3	7.3	6.3	7-6	6-4	6-2, 6-4
5.4	7.4, 7.1	6.4, 6.1	7-3	6-2, 6-3	6-2, 6-4
5.5	7.4, 8.1, 20.1	6.4, 7.1, 7.6	7-1, 8-1, 19-6	6-3, 7-3	Ch. 6 Introduction
5.6	8.2, 8.3, 8.6	7.2	8-1	7-4	6-4
5.7	7.4	6.4	7-3	6-3	6-1
5.8	7.5	6.5	7-7	6-5	6-9
5.9	8.4, 8.8	7.1	8-4	7-4	6-6, 6-8
5.10	n/a	n/a	n/a	n/a	n/a
6.1	n/a	n/a	n/a	n/a	n/a
6.2	9.1	8.1	9-4, 9-6	8-2	7-3
6.3	9.2	8.1	9-5	8-2	7-2
6.4	9.2	8.2	10-2	8-2	7-8
6.5	9.4	8.4	10-4	8-4	7-6

(*continued on next page . . .*)

Screen Number (CD-ROM)	Serway: *PSE, 4/e*	Serway: *POP, 2/e*	Halliday, Resnick, and Walker: *FOP, 5/e*	Young and Freedman: *UP, 9/e*	Tipler: *PSE, 3/e*
6.6	9.4	8.4	10-3	8-5	7-6
6.7	9.6	8.6	10-3	8-6	7-1, 7-2
6.8	9.7	8.7	9-5	8-6	7-2
6.9	n/a	8.8	n/a	8-7	n/a
7.1	n/a	n/a	n/a	n/a	n/a
7.2	10.1	10.1	11-2	9-2	8-1
7.3	10.4	10.4	11-6	9-5	8-3
7.4	10.5	10.11	11-6	9-4, 9-7	8-2, 8-4
7.5	10.7	10.5	11-8	9-4, 10-2	8-2
7.6	10.8	10.11	11-10	10-5	8-3
7.7	11.1	10.11	12-1	10-4	8-6
7.8	11.3	10.8	12-4	9-4, 10-7	8-5
7.9	11.5	10.9	12-8	10-7	8-5
7.10	n/a	n/a	n/a	n/a	n/a
8.1	n/a	n/a	n/a	n/a	n/a
8.2	13.1, 13.2	12.1	16-2	13-2, 13-3	Ch. 12 Introduction
8.3	13.1	12.1	16-2	13-3	12-1
8.4	16.1	Ch. 13 Introduction	17-1, 17-2	19-3	13-3
8.5	16.2	13.2	17-3	19-2	13-1
8.6	16.1, 16.3	13.1	17-4, 17-5	19-3, 19-4	13-3, 12-1
8.7	10.1, 13.5	10.1	11-2	9-2	n/a
8.8	16.7	13.4	17-4	19-4	12-1, 13-3
8.9	16.7	13.4	17-5	19-5, 19-6	13-2
8.10	13.2	12.2	16-3	13-3	12-1
8.11	13.4	12.4	16-6	13-6, 13-7	12-5
8.12	13.4	12.4	16-6	13-6, 13-7	12-5
8.13	n/a	n/a	n/a	n/a	n/a
8.14	n/a	12.6	16-9	n/a	Ch. 14 Essay: Seismic Waves
9.1	n/a	n/a	n/a	n/a	n/a

(continued on next page . . .)

Screen Number (CD-ROM)	Serway: PSE, 4/e	Serway: POP, 2/e	Halliday, Resnick, and Walker: FOP, 5/e	Young and Freedman: UP, 9/e	Tipler: PSE, 3/e
9.2	16.5, 17.1	13.6 +	17-6 +	19-5 +	13-2, 14-1
9.3	16.6	13.7	17-11	20-2	13-1, 14-8
9.4	n/a	n/a	n/a	n/a	n/a
9.5	16.8	13.8	17-7	19-8	13-4
9.6	16.4	13.5, 14.1	17-9, 17-8	20-2, 20-6	13-5, 13-7
9.7	16.4	13.5	17-9, 17-8	20-6	13-5, 13-7
9.8	18.2	14.2	17-11	20-3, 20-5	13-6
9.9	18.3, 18.5, 18.6	14.2	17-11	20-3, 20-5	13-6
9.10	18.4	12.6	16-9	20-7	12-8
9.11	18.4	12.6 +	16-9	20-7	12-8
9.12	n/a	12.6 +	n/a	13-9	13-6 +
10.1	n/a	n/a	n/a	n/a	n/a
10.2	Ch. 19 Introduction	18.6 +	20-1, 19-1	15-1, 16-1	Ch. 15 Introduction
10.3	20.1, 20.2	17.1, 16.1, 16.2	19-1, 19-2, 19-6, 19-7	15-2	15-1, 15-2, 16-1, 16-2
10.4	19.1	16.1	19-2	15-2	15-1
10.5	19.5	16.4	20-3	16-2	15-4
10.6	20.5	17.5	19-9	17-5	16-4
10.7	22.1	18.1	21-4	18-3	17-1
10.8	22.1	18.1	21-3	18-6	17-1
10.9	22.3	18.3	n/a	18-7	17-4
10.10	22.7, 22.9	18.6	21-2	18-9	17-6, 17-7
10.11	22.8	18.7, 18.8	21-1	18-10	16-5, 17-6, 17-7
10.12	n/a	n/a	n/a	n/a	n/a
11.1	n/a	n/a	n/a	n/a	n/a
11.2	23.1	19.2	22-2	22-2	18-1
11.3	23.2	19.3	22-3	22-4	18-2
11.4	23.3	19.4	22-4	22-5	18-3
11.5	23.4, 23.6	19.5, 19.6	23-2, 23-3	22-6, 22-8	18-4, 18-5
11.6	24.2	19.8	24-4, 24-7	23-4	19-2

(continued on next page . . .)

Appendix B

Screen Number (CD-ROM)	Serway: PSE, 4/e	Serway: POP, 2/e	Halliday, Resnick, and Walker: FOP, 5/e	Young and Freedman: UP, 9/e	Tipler: PSE, 3/e
11.7	24.3	19.9	24-7, 24-9	23-5	19-3
11.8	25.3	20.2	25-1	24-2, 24-3	20-1
11.9	25.1	20.1	25-2	24-2	19-1, 20-1, 20-5
11.10	25.8	Ch. 20 Introduction +	n/a	23-6	20-6
12.1	n/a	n/a	n/a	n/a	n/a
12.2	29.1	22.1, 22.2	32-1	28-2, 28-3, 28-4	Ch. 24 Introduction, 24-1
12.3	29.5	24.2	n/a	28-7	24-1, 24-2, 25-1
12.4	30.3	22.7	32-9	29-7	25-4
12.5	30.6, 30.7	24.2	32-2	28-7	26-2
12.6	31.4	23.2, 23.4	31-3	30-2	26-1
12.7	31.3	23.1, 23.3	31-3, 31-4	30-3, 30-4	26-2, 26-3
12.8	31.1	23.1	31-6	30-6	26-2
12.9	30.8	24.1	32-10	29-10	29-1
12.10	31.7	24.2	32-11	30-8	29-2
12.11	n/a	n/a	29-5	n/a	Ch. 26 Essay: The Aurora
13.1	n/a	n/a	n/a	n/a	n/a
13.2	27.1	21.1	27-2	26-2	22-1
13.3	27.2	21.2	27-4, 27-5	26-4	22-2
13.4	28.2, 28.3	21.7, 21.8	28-4	26-5, 27-2, 27-3	23-1
13.5	26.1	20.7	26-2	25-2, 25-3	Ch. 21 Introduction, 21-1
13.6	32.2	23.1	37-1	31-3	26-7, 26-8
13.7	32.5, 33.5	24.4	33-2	31-6	28-4, 28-5
13.8	n/a	n/a	n/a	n/a	n/a
14.1	n/a	n/a	n/a	n/a	n/a
14.2	35.3	25.1, 25.2	34-1, 34-2	35-1	Ch. 31 Introduction
14.3	35.4	25.3	34-7	34-3	30-3
14.4	35.4	25.3	36-2, 34-7	34-3	30-4, 31-3

(*continued on next page . . .*)

Screen Number (CD-ROM)	Serway: *PSE, 4/e*	Serway: *POP, 2/e*	Halliday, Resnick, and Walker: *FOP, 5/e*	Young and Freedman: *UP, 9/e*	Tipler: *PSE, 3/e*
14.5	35.7	25.6	34-8	34-4	30-4
14.6	36.1, 36.2	26.1, 26.2	35-1, 35-2, 35-3	35-2, 35-3	31-1, 31-2
14.7	36.1, 36.2	26.1, 26.2	35-2, 35-3	35-2, 35-3	31-1, 31-2
14.8	36.4	26.4	35-6	35-6	31-4
14.9	n/a	n/a	35-7	36-6	32-5

Note: + indicates greater coverage in CD-ROM than in textbook.

Appendix B